SMART CITIES
BREAKING THE POVERTY BARRIER

RAJENDRA JOSHI

INDIA • SINGAPORE • MALAYSIA

Notion Press

Old No. 38, New No. 6
McNichols Road, Chetpet
Chennai - 600 031

First Published by Notion Press 2019
Copyright © Rajendra Joshi 2019
All Rights Reserved.

ISBN 978-1-64733-528-1

This book has been published with all efforts taken to make the material error-free after the consent of the author. However, the author and the publisher do not assume and hereby disclaim any liability to any party for any loss, damage, or disruption caused by errors or omissions, whether such errors or omissions result from negligence, accident, or any other cause.

While every effort has been made to avoid any mistake or omission, this publication is being sold on the condition and understanding that neither the author nor the publishers or printers would be liable in any manner to any person by reason of any mistake or omission in this publication or for any action taken or omitted to be taken or advice rendered or accepted on the basis of this work. For any defect in printing or binding the publishers will be liable only to replace the defective copy by another copy of this work then available.

DEDICATION

This book is dedicated to the people and communities who have
reposed their faith in the work done by Saath
and to my parents.

COMMENTS

"Saath ensured communities understood not just their rights when it came to basic services but also their responsibilities. With three decades of experience in partnerships for equitable and rights-based urban development, Saath is well positioned not just to be a player, important as that is, but to also be a resource agency, a teacher and a guru, sharing its successes and failures to other institutions who are treading a similar path."

Mr Shankar Venkateswaran,
Former Chief,
TATA Sustainibilty Group and
former Country Head,
American India Foundation

"Saath's work with young people is a fascinating journey of an organization learning to work with communities and of providing leadership in a way that it never crosses the boundary of an enabler, while ensuring that the young people are able to channelize their confusions, anger and passion for change and better life into building pathways for the entire community towards a dignified life."

Amitabh Behar,
CEO, Oxfam India

"In the last thirty years, SAATH has taken up number of approaches of working with poor. The chapter on Basic Services and Housing describes SAATH's work on facilitating sanitation in a chawl, provision of basic services through the Slum Networking Project (SNP), improving access to formal affordable housing and the facilitating housing for migrants. Each of these experiments have lessons for urban India"

**Prof Chetan Vaidya,
former Director of the National Institute of Urban Affairs and School of Planning and Architecture, Delhi**

"This book not only highlights the good work done by Saath, but also provides food for thought in terms of what needs to be done to make our cities a much better place to live than what they are today. This book will certainly help inspire people to join NGOs in their own ways and help create an atmosphere for social change that will lead to a more inclusive growth."

**Mr Dilip Chenoy,
Secretary General,
Federation of Indian Chamber of Commerce and Industries**

"In its practical accomplishments and in its exemplary style of work SAATH has demonstrated that, despite the apprehensions, and occasional hostility, of various levels of government in India, NGOs have a distinctive and productive role to play in the development of the country and its people."

**Professor Howard Spodek,
Historian, Shrenik Lalbhai Chair
Professor of Management of Social Organizations**

"Saath's journey over the last 30 years has demonstrated that professionals who have no philanthropic background can initiate powerful institutions. Having been witness, accompanier, consultant, mentor, board member and now, Chair of Saath, I give the credit to Rajendra who committed Saath to become a public institution rather than a personal, private, or family enterprise, which many non-profits have been reduced to in the country."

**Mr Gagan Sethi,
Chairperson, Saath Board of Trustees**

"Saath's three-decade journey is very interesting to read because it helps us understand the various hurdles one encounters while organizing the poor, especially when it concerns financial services."

Vijayalakshmi Das,
Managing Director,
Ananya Finance for Inclusive Growth Pvt. Ltd.
and Ananya Vijayalakshmi Finance

"Saath has done commendable work in the area of children's education and women and health and should be congratulated on several counts."

Ms Indu Capoor,
Founder Director, Chetna

"If Saath can power Urban India as a mere NGO, imagine how much more they could contribute to the national economy if empowered?"

Prof Ram Kumar,
IIMA alumni

"Saath can help in developing awareness, conducting training and play the role of a proactive disaster management institution, not only for Gujarat, but for the entire country."

Prof. (Dr.) C N Ray,
former professor at
the Faculty of Planning and Public Policy,
CEPT University, Ahmedabad

Contents

Foreword *11*
Preface *17*
Acknowledgements *23*

PART 1

Chapter 1:   India's Forgotten Citizens 27
Chapter 2:   How to Break Through the Poverty Web 36
Chapter 3:   The Revolutionary Initiative 40
Chapter 4:   A Roof above their Heads 54
Chapter 5:   Livelihoods – Teaching Communities How to Fish . . 70
Chapter 6:   Financial Inclusion 98
Chapter 7:   How to Channelize and Empower People 115
Chapter 8:   Bringing Women to the Fore and
             Preventive Health 124
Chapter 9:   Children and Education 136
Chapter 10: Harnessing the Power of the Youth 153
Chapter 11: Market Engagement 165
Chapter 12: When Disasters Strike 171

PART 2

Chapter 13: Organisation Development and Governance 205
Chapter 14: Partnerships 215
Chapter 15: Social Impact 223
Chapter 16: People at Saath 229
Chapter 17: My Personal Journey 269

Notes from Experts *275*

ANNEXURES

*Annexure 1: Social Return on Investment
Study on Saath conducted by 4th Wheel* *307*
Annexure 2: Reconciliation Study after Gujarat Riots (2004) *323*
Case Studies *343*
*Annexure 3: Review of Saath's Urban Initiatives by
Shrawan Kumar Acharya, PhD in September 2007*. . . *349*
Saath's Urban Initatives *351*
Bibliography *383*
Appendix I: Diversity Chart *385*
Appendix II: Saath Coverage in Ahmedabad City *387*

FOREWORD

by Mr Shankar Venkateswaran
**Former Chief of Tata Sustainability Group,
Former Country Head, American India Foundation**

When I was asked to write a foreword to Saath's publication describing its seminal work of over 3 decades, I was delighted. The organisation I was then with – American India Foundation (AIF) – was familiar with Saath's work in the wake of the Kutch earthquake, but the two organisations began working together in 2002 post the carnage in Gujarat. This partnership continued for several years, especially in the skilling space. My personal relationship with Saath and its founder, Rajendra Joshi, continued well after I left AIF in 2008 and, as it turns out, I have known the organisation for nearly half its existence!

Thirty years in development is no mean feat and Saath is certainly entitled to look back with pride on what it has achieved, which, as you will gather from the book, is significant. Numbers are often the tip of the iceberg and more details can be found in the book: touching lives of over 1.5 million people, some 106,000 women and children supported for immunisation, safe delivery and nutrition, enabling over 62,000 with jobs, livelihoods, entrepreneurship, assisting over 4,000 families and so on. And, of course, the book also talks about all its achievements that are hard to quantify but are perhaps more significant, such as the community institutions it has helped foster and community leaders like Devuben (someone who I admire greatly) who have emerged from its work.

But this book is much more than a chronicle that celebrates achievements. It is full of reflections and lessons that any serious student or practitioner of poor-centred urban development will find extremely instructive. It is also, remarkably, a story of transformation – of a development approach implemented through a set of projects to creating an institution. Like many, I will always be an admirer of Saath's programmatic interventions but I think the vision to build an institution that outlasts its founder – and succeeding to a great measure – is what legends are made of. Those who want to know more about this aspect should not only read this book but also spend time with the architects of this transformation – Rajendra Joshi himself, Gagan Sethi, and all the colleagues who have been a part of Saath during this journey.

Before Saath, my exposure to development was essentially rural. I had "defected" from the corporate sector in 1995 and had the opportunity to see and learn from the work of some of the finest grassroots NGOs across India. However, all of this was rural, and when the organisation I was with at the time began to address urban development issues, we found that urban challenges were so different that we had to "unlearn" our rural experiences and build partnerships with a completely new set of urban NGOs. Saath was the first NGO I was associated with whose origins and experience were essentially urban. It had built an envious track record already. It became my urban development guru.

Urban development in India always presented a different challenge than the more familiar rural development and this was even more so in the 1990s and early 2000s when Saath came on the scene. Caste and gender remain the dominant bases for exclusion and injustice, but these are less visible and geographically harder to pinpoint in urban areas. Urban livelihoods (unlike their rural counterparts) are in the form of jobs (mostly informal as far as the poor are concerned) and in the services sector rather than production, and even these are diverse and continuously evolving. Thanks to the messy process of urbanisation, homelessness and haphazardly built slums with no basic services such as drinking water, sanitation and hygiene is sadly a very common and visible urban phenomenon. Women, as always, are worse off with no sense of privacy to have a bath or use the toilet. Access to basic services is often determined by politics than on the basis of human rights and fighting for these entitlements is fraught with physical danger of eviction. Challenges in education and health have less to do with infrastructure and the presence of trained teachers, doctors, and nurses at schools and hospitals, and more with finding physical spaces to locate them.

Unlike village panchayats, there was really no urban governance tradition in India. As a result, Indian towns and cities are in reality not governed as much as administered by a municipality run by a bureaucracy that do not see poor communities as partners. The municipality is responsible for providing basic urban services like water, sewage, sanitation, public transport, and dealing with solid waste and, in some cases, electricity distribution. It maintains all infrastructure and runs schools and health facilities (parallelly, the state and private actors are also present). Importantly, from the viewpoint of the urban poor who live in the slums, it determines which slums are to be "recognised" (and hence provided with these basic services) and which should be razed to the ground, based essentially on politics. Thus, any NGO working with the urban poor has no choice but to work in partnership with the municipality, which is usually not similarly disposed.

It is in this context that Saath's work needs to be seen. To me, its biggest advantage was that it was not an NGO that "migrated" from the rural to the urban, like many I had encountered. It did not have any "baggage" or set way of working, which it had to then adapt to an urban context; it learnt development in urban India, experimented, and built its own pathways. Its second advantage was that it was Gujarati – in origin and where it worked. I am not alone in the belief that there is something in the air in Gujarat that makes its people and institutions entrepreneurial and innovative, constantly striving towards efficiency and productivity – the "dhandho" spirit that focuses on output and outcomes. Saath as an institution has that in immense measure. And the third was its leadership, sheer competence and its core value of working in partnership – by the time I got to know Saath, this was already a part of its DNA, built on three pillars – Acceptance of different points of view, Respect and Trust – 'ART' for short.

Saath realised early on that if the urban poor has to be served, it must work closely with the Ahmedabad Municipal Corporation (AMC). And of course, communities must be central to this partnership. Together they have addressed issues of water, sanitation, drainage, street lighting and housing in slums. Saath's role as a facilitator and interlocutor between AMC and the communities has been one of its significant contributions, an early example of community participation in an urban development scenario. **Saath ensured communities understood not just their rights when it came to basic services but also their responsibilities.** AMC too recognised this role of communities, leading to engagement with dignity and sharing of responsibilities in planning, financing, execution and maintenance, a far cry from the top-down approach of most municipalities and, perhaps, also demonstrating the "dhandho" spirit!

What has always struck me about Saath's approach to development is that it is firmly rooted in reality and pragmatism. Its work with Home Managers is a case in point. The relationship between a housemaid and her employer is informal and hence often undignified and exploitative, but one where there are benefits for both. Saath first went about upgrading the skills of housemaids so that they became Home Managers; this then made it easier to formalise the relationship between her and her employer where wages, hours of work, holidays etc. were agreed. Since upgrading skills was not Saath's core competence, it built partnerships – with hotels, appliance manufacturers, and so on.

I referred earlier to the great institution-building project that Saath has accomplished, involving governance, staff engagement, succession planning, and a host of complex issues that organisations have to grapple with and where many NGOs stumble. I will leave it to you to read the relevant chapter and savour it. However, I must relate a specific instance which I believe illustrates the sophistication of Saath's thought process when it comes to its work and its staff. It relates to one of the most polarising events in the history of independent India – the Gujarat riots of 2002. Saath continued to not just provide succour to the victims but also to rebuild trust between the communities and this book describes some of the humongous challenges they faced and overcame. The staff at Saath was not untouched either. I remember Rajendra mentioning one particular staff member who believed that "they" deserved what happened to them. The Saath team could have taken the easy way out and simply asked him to leave. But they did not. The challenge was not that this individual was causing disruption, they thought, but the mindset behind the belief that "they" deserved it. So, what did they do? They kept him on and invested in trying to get him to reflect on his beliefs and change. To do otherwise would have begged the question – can you build harmony between communities with a team that does not share the same harmony?

What does the future of urban development hold? It is quite clear that urbanisation will continue apace and continue the way it has in the past; the urban poor will be the greatest sufferers. Having said that, it is also clear urban planning and management have become extremely complex and no one institution – state, private sector or civil society – can be expected to find all the solutions. There is also a growing consensus that in the future, innovation will not come from competition, but through collaboration. Thus, partnerships are the only way forward and this has been clearly articulated as Sustainable Development Goal 17. With three decades of experience in partnerships for equitable and rights-based urban development, Saath is well positioned not just to be a player, important as that is, but to also be a resource agency, a teacher, and a guru, sharing its successes and failures to other institutions – government and non-government – who are treading a similar path.

PREFACE

I live a fairly comfortable life in Ahmedabad. My day starts in a relaxed manner – breakfast and reading the newspapers. This remains a fixed event in my life, and, when I think about it, it is possible because the milk for my tea and newspapers are delivered like clockwork every day. Then my breakfast plates and tea cups – and of course the innumerable other dishes used through the day – are washed and sorted by our domestic maid, who also does the laundry, sweeping, and mopping. As a result, I live in a neat and clean home, without having to do any of the chores. At night, I know I am safe because of the watchmen at the gated community where I live. I get to my office on time and less stressed than most other commuters who are dependent on public transport, because I have a driver who navigates through the chaotic peak hour traffic and spares me the trouble. Once again, at the office, everything is clean and in its place. The cleaning staff have done their job. So, all I have to do is get straight down to work, to earn my living. The office boy keeps me hydrated and well-fed with beverages and snacks, whenever requested.

But this is not all. I have other comforts that make my life easy.

I have the service of auto rickshaw-wallas, vegetable and fruit vendors, dhobis, dog-walkers, postmen, etc. who ensure that my daily essential needs are met. If I lived in Mumbai, a home-cooked lunch could also be delivered by a dabbawalla.

If I were in the construction industry business, masons, plumbers, carpenters, fabricators and electricians would be the mainstay of my trade. If it were in the automobile industry, I would have mechanics, factory workers, and cleaners. If in the IT industry, I would employ data entry and call centre persons. The list is endless.

Every day, we use the services of a whole host of people, who we never stop to think about. Perhaps, we pay them their salaries and give him other benefits. But how often do we stop to think about how *they* live? What makes *their* days easier, seamless, safe? How often do we think about what makes *us* and *them*?

The dabbawallas, carpenters, domestic help, sanitation workers, food vendors, the "informal sector" per se also have families, homes, children to educate, elderly to be taken care of, run their small businesses and earn a living.

How do they cope? Do they have social security, bank accounts, credit cards, insurance – health or otherwise, that the middle class have? Do they even have basic amenities like running water and sanitation?

Most of us interact with these people every day. Do we recognize their considerable contribution in making our individual, family, work, and business lives comfortable and efficient? How many of us appreciate their role in the success of our businesses? Do we acknowledge their impact in the sustenance and growth of our cities and towns? Do we understand their aspirations and challenges? Do we know where and how they live? Are they compensated adequately?

Residential areas in urban India are broadly divided on the basis of income levels and livelihood of its occupants. The middle and higher classes, with higher incomes, mostly employed in the formal sector, live in formal housing by way of flats, bungalows, etc.; while the lower classes, with lower incomes, mostly engaged with the informal sector are consigned to live in chawls, slums, government housing, and low-cost housing. Irrespective of where they live, there is a symbiotic economic and social relationship between the rich, the middle class, and the lower classes.

How has this highly inequitable situation come about? At a fundamental level, it is a failure on the part of our planners and policy makers to create an appropriate infrastructure for secure and hygienic shelter for a significant portion of the city's population.

Most in this "Informal Sector" live in squalid conditions; and they are the important support structures in our "Smart Cities." According to the 2011 Census, there were 13.92 slum households with a population of 65.49 million, which was 5.41% of India's population. In August 2013, the government of India, in parliament, stated that the slum population was projected to reach 104 million by 2017, which would be 9% of the projected total population.

The population/land occupied ratio of slums and chawls may come as a surprise. In Mumbai, 55% (6.5 million people) of its population lives in slums and occupies just 12% of its land mass. In Ahmedabad, slums host a population of 728,000 (13.1% of total population and occupy just 1.16% of its land mass.

These high population density figures indicate that slums provide an efficient housing option for a large proportion of population in our cities. However, many city planners and governments label slum areas as illegal and want to evict them. A good question to ask is what is a viable alternative for housing these people and at what cost?

India is undergoing rapid urbanization. According to 1901 census, the urban population in India was 11.4%. This increased to 28.53% according to 2001 census and increased to 31.16% and projected to be 33.54% in 2017.

India has launched an ambitious Smart City initiative in 2015 to address urbanization. The mission of the Smart Cities Project is to drive economic growth and improve the quality of life of people, by enabling local area development and harnessing technology, especially technology that leads to smart outcomes.

There is a lot of debate about what constitutes a smart city. But what are Smart Cities?

Are they cities, which function because of digitization and technology? Cities, which attract investments in industries? Cities where citizens, especially women and children, live without fear? Cities where commuting by public transport is the first choice? Cities where all citizens have access to decent basic services? Cities, which are culturally active? Cities where the gap between the haves and have-nots is decreasing? Are our cities inclusive? Do our cities provide opportunities for decent livelihoods and jobs?

There is no specific answer because different industries will answer this these questions through their own biases. For example, an IT company will call a city smart if, let's say, there is free wifi in all public transport.

At Saath, we too have an answer. We believe that cities will become smarter when there is **equitable sharing of resources, inclusion of informal sector populations, and social cohesion**.

Urbanization is driven by migration. Migrants come to cities with abundant aspirations and energy. The USA is a country built by migrants, who were welcomed and whose energies and aspirations were channelized for economic development. Opportunities were created for migrants so that they could learn skills, get jobs, educate their children, were given social security, and housing. Policies and programmes enabled a seamless mainstreaming into society. Are our cities able to harness the energies effectively for economic growth? We do not think so, because a migrant faces many hurdles, by way of documentation and legal status, housing and livelihoods.

Globally, there is recognition that slums and low-income neighbourhoods can be a driving force for the growth of a city. Instead of destroying slums, the way forward is to mainstream slums, where residents have created rich, social, and economic networks, into the fabric of the city. The synergies between the informal and formal sectors can make a city thrive and become self-sufficient.

At Saath, we do not accept a situation in which people in the informal sector, who contribute significantly to the growth and sustenance of our cities, be consigned to live in unhygienic slums and chawls.

In this book, I will show you all that I have learned and experienced working with individuals and communities residing in slums and chawls.

The elements that sustain these large populations. Where they come from and why they migrated. How they manage their lives. How they have been neglected in the planning and development process of our city. How the political system protects them and also uses them. How they help our cities work more efficiently. How concerned people and organizations are working towards making a difference.

In popular mythology, slums are supposed to be dens of crime with high alcohol and drugs consumption. My experience is that this myth is far removed from the truth. Slums have crime rates, and drug and alcohol usage which are similar to the rest of the city.

I have attempted to highlight the challenges, innovations, and solutions that have characterized migration and urbanization in India, as well as the roles played by migrants, entrepreneurs, governments, civil society, and corporates in facilitating the process of urbanization.

As per estimates of the Committee set up by Ministry of Housing and Urban Poverty Alleviation under the Chairmanship of Dr. Pranob Sen, Principal Adviser, Planning Commission, the slum population in the country was expected to touch 93.06 million by 2011. At a per capita of USD 1,504, this is almost a USD 132 billion economy. Is the contribution of migrants and the informal economy, which contributes to the growth of our cities, recognized?

When cities are struck by natural and man-made disasters, vulnerable populations are the most affected. Saath's work in this area has been extensive. We worked tirelessly with the affected populations during the Gujarat earthquake in 2001 and, the Gujarat Riots in 2002. The earthquake took us to the rural Khadir area and Rapar town in Kutchchh. The riots affected the slum dwellers immensely. Rehabilitation of both these disasters was a mix of assistance, teaching, as well as learning. Our work with rehabilitation of communities affected by the laying of four-lane highways and the Metro Project in Ahmedabad was again, a different kettle of fish, dealing with a totally different aspect of rehabilitation.

This is not a treatise on urbanization. It is about how the unorganized sector, living in low-income neighbourhoods, makes a city smart; and how, we at Saath, have worked to enhance synergies between various stakeholders to help make our cities smart.

I will take the reader through Saath's journey; its evolution as an organization with multiple entities, that provides a non-threatening and reassuring platform for people and organizations that want to make our cities better places to live in.

I hope this book inspires and lights the way for others to take up community initiatives such as Saath's to change, empower and recharge our country's forgotten citizens. If we can do it, so can you.

Acknowledgements

The success and growth of any organisation depends on multiple internal and external factors. At Saath, we have been very fortunate. We have had a motivated and committed team. The communities that we have worked with have been trusting and supportive. Our partners have had faith in us that we will deliver as promised. Committed individuals have spared valuable time and efforts to serve in our Board of Trustees and Audit Committees. We are grateful to all these stakeholders for helping Saath in serving vulnerable communities in this journey of thirty years

For creating this book, we are grateful to Mr Shanker Venkateswaran for writing the foreword and Mr Mr Gagan Sethi, Mr Dilip Chenoy, Prof C. N. Ray, Ms Indu Capoor, Prof Chetan Vaidya, Ms Vijayalaxmi Das, Prof Amita Bhide, Prof Ram Kumar and Mr Amitabh Behar for reviewing various chapters of the book and sharing thir comments and insights.

We are indebted to Saath team members and people from the communities had not taken out time to share their memories and experiences. Without them this book would have been difficult to write

We are thankful to Ms Vanessa Nazareth who converted raw writing into readable material and Ms Trisha Bora for making the book engrossing for its readers.

PART 1

CHAPTER 1

INDIA'S FORGOTTEN CITIZENS

Our horror is their daily life

When Ramjibhai came for the interview, he was nervous. He thought he was going to be asked to give a speech. There was a palpable look of relief on his face when we told him that it was just a one-on-one interview.

Ramjibhai is sixty-three years old and has lived with his family in Sankalchand Mukhi ni Chali (SMC) for the last forty-five years, where his father had been allotted a house by the textile mill he had worked at. He is from Saurashtra, and he belongs to the Vanker Dalit community. Ramjibhai now works as a security guard. Earlier he had worked as a painter in the Diamond Silk Mill located in the Thakkerbapanagar area. Like many other textile workers, he lost his job when the factory closed down during the collapse of the textile mills in Ahmedabad.

Ramjibhai's interview was part of our first engagement with slums in 1989. Our survey focused on the residents of Sankalchand Mukhi ni Chali (SMC) in the Behrampura ward of Ahmedabad in Gujarat. SMC was originally built in 1920 for housing about 300 textile worker's families. As the textile industry declined in Ahmedabad, the owners stopped maintaining the chawl, but still collected rent. By 1989, it was overcrowded with about 500 families and the water and drainage infrastructure was highly stressed. Our socio-economic survey brought to light the dismal state of the chawl. What we found was shocking.

Sanitation was in an abysmal state. Incidence of tuberculosis was alarming. Financial inclusion and access to formal credit was almost absent and the dropout rate among children was high. Ramjibhai's account will tell you more than any of our reports ever can. We asked him about his living conditions.

"The sanitary and hygienic conditions were terrible in SMC earlier," he said. "There were only 14 toilets in the chawl which had about 300 houses," he said. "People were forced to defecate in the open and this led to disease. The old and sick were at a terrible disadvantage. There were fights among residents in the long queues."

If you've ever taken and early morning bus or train going into an Indian metro, you would've probably noticed people defecating in the open. In Mumbai, all across the city, it is a common sight to see hordes of people doing their morning business by the sea-side. Most of us look away in disgust or laugh at the sight. "Look at these people, they have TVs and refrigerators in their houses, but they don't build a toilet. How shameful!"

But our horror is their daily life.

And it's because they have no other choice. The truth is much more complicated and dismal. What most people don't understand is that a person living in a slum can go and buy a TV or a fridge in a shop. But there is no way he can buy a toilet in a shop! A toilet requires drainage and a water connection, which requires permission from civic authorities. Both these are not available to the slum residents. Defecating in the open is not a choice they would make, given an option; but the unfortunate truth is that, they have no option.

This is just one example of how slums and slum dwellers are at a major disadvantage at being left out of the formal organized sector. Smart cities are those that account of every life that lives in it. Not just the ones who can afford it. Before I go onto to show you how we can make our cities more inclusive and efficient, we must understand how poverty and slums function. To understand the bigger picture, we must first get to the root.

How slums are created

The creation of slums can be directly attributed to a lack of foresight on the part of our planners and policy makers. Post-Independence, India's policy makers chalked out a path of economic growth that emphasized industrialization. Subsequently, India witnessed a shift in labour markets from the agriculture sector to the industrial and services sector, which for economies of scale are located in or near urban areas. What our policy makers did not plan for was the housing for people who would migrate to urban areas to provide labour for the industries. In effect, migrants were left to their own devices to find affordable housing and accommodation. Slums became the default affordable housing for a majority of migrants, because formal housing options were limited and expensive.

A common myth is that slums are created by people encroaching and occupying empty land. The truth is more complicated. Since our city planners and authorities do not plan housing for migrants, the market provides a solution. Every town and city in India has land, which can be termed "grey land"; this "grey land" roughly categorized are: hazardous land, land, which has disputed land titles and unoccupied government land. Hazardous land comprises of river banks, landfill sites, and land near railway tracks and highways. Land with disputed titles are either common land of villages, which have been incorporated into cities and land which came under dispute after the Urban Land Ceiling Act of 1977.

In the early 1990's, chawls had been built by industries to house their workers. When these industries closed down, investment for maintenance of the chawls stopped and they became decrepit. The residents in the chawls had disputes with the industries regarding their termination and continued living in the chawls. To compensate for their loss of income, residents stated giving parts of their homes on rent. Gradually, chawls became congested and slum-like.

These lands and chawls constitute the grey land market. Estate agents in the informal economy purchase these lands with the connivance of officials and land owners. They then parcel the land into smaller plots and sell these to migrants or to middlemen who create basic housing structures and then rent them to migrants. As a result, a huge grey and informal housing market is created in which houses are sold, bought and rented. This grey market has its informal, but not legal, documents, which are tradable and serve as proof of ownership and occupancy. These are our present-day slums and chawls. It is estimated that there are approximately 1.35 crore households valued at approximately Rs. 13,500 crores built privately, at no cost to the government.

The failure of city authorities and planners to create a formal, market-based mechanism for providing affordable housing for the millions of people who migrate into our cities, effectively create slums.

Since the city authorities and the middle class are hostile towards slums, residents look upon the political class for protection and a relationship of mutual dependence is created. Slum residents feel safe and may even get public sanitation and running water facilities and, the politicians a readymade vote bank.

As a majority of slums are termed "illegal" due to the absence of formal legal titles, city authorities do not provide basic services of water supply, sanitation, drainage and garbage collection. Without basic amenities, these settlements become unhygienic and lead to a sub-standard quality of life.

The everyday life of these "secondary citizens"

Saath's first job at hand was to deep dive into the lives of slum residents to understand their quality of life and aspirations. We had discussions with individuals, families, and leaders in slum and chawl communities. We conducted exhaustive socio-economic surveys and analyzed information and learnings. What we found was shocking.

Slum dwellers live in a constant fear of being evicted. In most cases, they have paid middlemen for land, which does not have a legal title. The city does not provide affordable housing with legal land title. The slum dwellers cannot leverage any credit by using the house or land as acceptable collateral. They cannot pay taxes even if they want to, because taxation gives them legal status. In short, lack of legal land tenure convert slum and chawl dwellers into secondary citizens of the city.

Slums do not have basic services. The ambiguity of land title and legal status prevents provision of formal infrastructure. As a result, political patronage becomes an alternative for planned development. Even when they want to pay for these services, they are unable to get water, drainage, or sanitation. They have to make do with public toilets and water supply with all associated problems. Even when they keep their houses clean, they do not have option for solid waste removal from the slum settlement.

A majority of the slum residents are first generation migrants from the rural areas. They possess skills that do not have any significant market in the cities. They have to learn new skills, which require long-term investment of time and resources, which they do not have. Their only option remains is getting engaged in occupations that are unskilled and menial, and where the incomes are low. These are also casual jobs. Irregularity of employment and no proof of address disqualify most of them from accessing credit for small businesses. Without a land title for the house, they are unable to offer collateral for any credit. Unscrupulous companies also take advantage of the ignorance of slum residents and cheat them. Banking in the formal sector does not welcome disbursement of small loan amounts to the slum residents, as the procedure is quite expensive for these organizations. Hence, the slum residents are forced to borrow from usurious moneylenders.

Majority of children in slums are first generation students, whose parents may be illiterate. They do not have parental support or guidance required for coping with formal education. They lack learning materials. Moreover, children from slums are not "properly" clothed. Shortage of water prevents cleanliness. These factors, when compared to those of other children attending school, leads to a condescending attitude from teachers who label them as 'underachievers.'

Children from slums are looked down upon. These reasons lead to a high failure and dropout rate. For girls, formal schooling is more difficult because traditional attitudes do not favour long-term education. Even the few who manage to secure a college degree are disillusioned when they fail to get employment. They then have to learn a new earning skill. This leads to a belief that 12 or 15 years of formal education are a waste of resources. Lack of formal education closes opportunities for technical education, making learning of formal earning skills difficult. Consequently, a majority is forced to join the informal sector, doing menial work.

Health is a major economic issue for slum residents. The unhealthy physical environment leads to sickness, which leads to costs for continuing medical treatment, which leads to reduction of workdays and economic loss. Economic loss leads to inability to invest in clean environment. The vicious cycle continues. Treatment at government hospitals is apparently cheaper but is inconvenient to the slum residents because of time lost in waiting for the treatment and often, indifferent attitude of the medical staff. Instead, they prefer more expensive private treatment.

Ignorance and lack of proper education lead to continuation of wrong beliefs and an unscientific attitude towards health. The outcome is incomplete immunization, insufficient gynecological checkup during pregnancy, unsafe deliveries at home and improper post-natal care of mothers and children, especially in terms of diet and immunization. Incomplete tuberculosis (TB) and malaria treatment lead to recurrences and relapses. The need for fast cures helps propagate the myth that expensive treatment is good treatment.

Women and girls in slums have a poor quality of life. The lack of basic services affects them the most. They have to spend considerable time collecting potable water and getting rid of wastewater. Having to defecate in open spaces is a health and social hazard. Women are most disadvantaged in slums; they have to look after children who are frequently sick, husbands who do not earn adequately and, are often drunk, as well as try to ensure that the family gets a meal every day. Girls have to look after younger siblings when both parents go to work. Combined with a traditional bias against educating girls, they are often not sent to school or drop out at an early stage. Girls do not have the exposure to everyday city life situations, which men, women and young men have. As a result, they are often anxiety prone and stressed.

The unhealthy and polluted environment, lack of immunization, malnutrition, and absence of educational exposure affects children in slums. Sadly, their physical, emotional and intellectual growth is stunted from a very early age.

Men in slums have inadequate earning skills leading to low incomes and an inability to provide adequate resources for household expenses. This leads to frustration, which is often expressed through escapism in addiction of various types. The need to provide an adequate quality of life combined with a lack of formal earning skills forces the option of extra-legal activities. The youth are frustrated, as they do not have the opportunities that their better-off peers have. This results in a cynical and diffident attitude, which becomes a handicap during adulthood.

But it need not be this way. Saath found a way to break through the web of poverty and empower the residents of SMC in 1989.

How we broke through the poverty chain

In Sankalchand Mukhi ni Chali, there were 500 households with 20 semi-functional, filthy, occasionally maintained, public toilets to cater to their sanitation needs. Our socio-economic survey indicated that household toilets were a MAJOR priority. With active young men of the slum, who became social animators, we found a way to construct toilets at a household level.

The Ahmedabad Municipal Corporation (AMC) had a programme for construction of toilets called the 80:20 scheme. In this scheme, AMC would give an eighty percent subsidy for construction of a toilet. A water connection would also be provided. The main hurdle was that the resident had to invest about Rs. 4,000 to construct a toilet. It had to be according to the guidelines set. Once completed, it had to be approved, and then the 80 percent subsidy would be reimbursed. Most residents could ill afford the initial Rs 4,000 investment and did not know how to go about acquiring it.

Saath, along with a youth group, studied and understood the process of applying for the scheme and then approached Oxfam for a revolving loan to construct 20 toilets. Criteria for disbursing the loans, included, ability to repay; priority was to be given to pregnant women, the sick, and the elderly. The programme required personal bank accounts, which we facilitated. This was the beginning of financial inclusion for the communities we worked with. Loans were disbursed to 20 households. Contractors who could build according to specifications of the scheme were identified.

Construction of toilets commenced.

Youth groups along with the residents supervised the construction. Payments were made to the contractors based on phases of construction. Within three months, the first lot of 20 toilets was completed. Approvals were sought and given by the AMC, who then reimbursed the subsidy. Simultaneously, the youth group identified the next batch of 20 residents and the process was repeated. By 1991, almost all houses had a toilet with a functioning water connection, which they maintained well. These households also benefitted with a functioning bank account.

Coming back to Ramjibhai.

It was not long before he heard about the toilet construction programme that Saath and the youth of SMC were promoting. "I got the details and filled out the application form," he said. "I got a loan of Rs 2,400 and paid Rs 800 as my share."

When asked whether paying his share was a challenge, he said. "Yes, it was. But it was such a necessity that I did not hesitate."

When his application was approved, he kept close watch while the toilet was being built.

"I was investing my hard-earned money and wanted to make sure that the toilet was built as per specifications," he explained. "It was a big relief to the family when the toilet was built. We could use it whenever we wanted. We did not have to stand in a queue or endure any discomfort," he said. **When asked whether he had approached local authorities earlier, he replied in the affirmative, adding,** "The local corporators and officials did not care. It was better paying for the toilet than waiting indefinitely."

This was our first Integrated Slum Development (ISD) project with government cooperation, the community, and NGO collaboration and it was successfully implemented! This initiative enabled the community to construct toilets at affordable rates, the AMC to meet its target of promoting construction of toilets, our objective of improving the quality of life, and Oxfam's objectives of assisting the poor; a win-win situation for all partners. This experience set the tone for future initiatives for facilitating basic services. At the same time, the youth of the slum realized that they could bring about positive change. They became more self-confident and were recognized as leaders by the community. The community realized paying for services reduced their dependence on middlemen.

Eventually, all the houses in SMC had a toilet. The chawl became much cleaner.

"We could invite our relatives over to visit us without feeling ashamed," Ramjibhai said. "People started extending their houses as their families grew, and the chawl is much more crowded now. I regularly pay the Rs. 10 monthly rent." he said proudly.

Next time, you see a slum, remember that people living there have no other option.

They do not have access, like you and me, to either legal affordable housing or housing finance. The authorities term them "illegal" and do not provide water or drainage. They are left to their own devices to figure out their shelter requirements. They are at the mercy of the hostile urban elements.

Basic services of water, sanitation, and drainage, as well as decent shelter, are a prerequisite to a better quality of life. Ramjibhai's story is one of the many lives Saath's initiatives has touched and improved.

In the next chapter, I'll show you through a comprehensive plan how we as a community can reverse and empower people who have been swept under the carpet by the formal, organized sector.

Chapter 2

How to Break Through the Poverty Web

The quality of life of any person depends on access to shelter, livelihoods, health and education services, financial inclusion, and an ability to negotiate with authorities and institutions providing these services. These aspects, which determine the quality of life, are dependent on each other. Inadequate sanitation, drainage, and water supply lead to sickness, which affects earning ability. Lack of adequate income leads to inadequate spending on health services and education. A house without electricity negatively effects children's ability to study and subsequently leads to an occupation with low income. Lack of access to financial services and loans lead to borrowing at high interest rates, which, in turn, leads to inadequate spending on health and education. Lack of residence proof means that bank accounts cannot be opened and thus no access to government schemes.

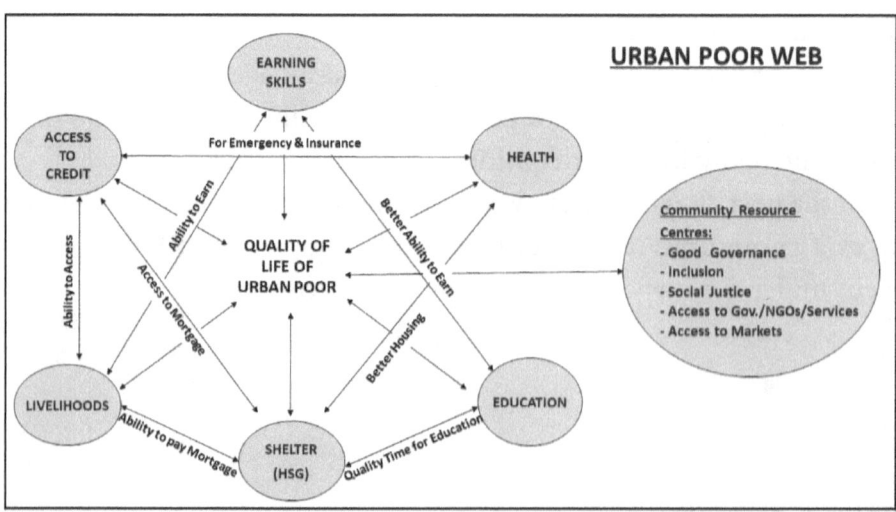

Saath's analysis was that, unlike middle class families living in formal housing with stable incomes, people in low income neighbourhoods are caught in a web of poverty, and that addressing any single aspect of quality of life will not lead to significant change. Improving education of children will not make a significant change in shelter. Enhancing only livelihoods will not result in better sanitation and water supply. Only financial inclusion will not lead to better health.

We designed an integrated approach to address all the aspects of the poverty trap simultaneously, in which, partnerships with all stakeholders would be paramount. We called this approach Integrated Slum Development.

The Integrated Slum Development (ISD) Approach

Inherent in the ISD approach was to empower slum residents by making them participants in their own development. This was a significant departure from the welfare approach of central planning by the state till the 1990s, which had created an overall feeling that slum residents were inferior citizens of the city and hence the government had to take the burden of developing the slums. From an activist's viewpoint, there was a greater concern; that slum residents themselves were internalizing this attitude, leading to decreased self-esteem and a diffident attitude. The ISD approach assumes that the slum residents themselves were willing to be active participants in their own development process.

ISD sought to create opportunities through which slum residents became active change agents of development. It was thought that meeting the basic developmental needs of health, education, economic betterment, and physical services in a tangible manner could create the opportunities/options for slum dwellers. It would enhance their quality of life and in this proactive process bring about change and enhance the self-image of slum residents.

Activities and programs of the ISD framework are designed to enable slum residents to realize their potential by increasing their management and technical expertise, nurturing leadership, and enhancing their self-esteem and self-confidence. These abilities would start a virtuous dynamic cycle of actions for development. The intervention would begin with improvement in skills and capacities, that would lead to improvement in quality of life, that would enhance their self-worth, which would lead to critical reflection and finally, to more actions for development.

The effects of ISD are not restricted to low income neighbourhoods. The whole city gains when there is equity and social cohesion amongst its citizens. The components of ISD to catalyze change are:

- Women, children, youth, elders, people with disabilities, migrants and informal sector workers are primary participants who decide, pay for, and implement programs
- Design and implement programs that impact livelihoods, food, shelter, health, education, financial inclusion, social security and environment sustainability
- Saath's role, collecting information and data, facilitating and implementing leading to awareness, advocacy and policy change, and delivery of services
- Alignment with markets is fundamental in program design
- Partnerships with government, civil society, academic institutions and private sector a vital part of any program.
- Values of empathy, inclusiveness, transparency, collaboration, nurturing is inherent

ISD is the vision which has driven Saath. ISD provides a platform for all concerned stakeholders to participate in making our cities inclusive and smart.

How ISD works?

The role of Saath in ISD is that of a catalyst and facilitator. As a catalyst, Saath identifies a need or concern that can enhance the quality of life of the slum residents and design a response or a solution with communities. As a facilitator, Saath identifies stakeholders who, apart from the community, can be government agencies, civil society, private sector, academia, and experts. Saath then structures a programme with appropriate stakeholders and then implements the programme with communities. The programmes are designed such that a win-win situation is created for all stakeholders.

Saath provides an inclusive and nurturing platform called the Urban Resource Centers (URCs) for individuals, communities, and organizations that want to bring about societal change. Individuals can be staff, community members, students, and professionals. With communities and organizations, this platform provides an opportunity to engage in development activities to help meet their objectives.

At Saath, people are central in the ISD approach. The core belief is that an institution is what its people and their values are. The pedagogy approach of Saath is for individuals to recognise their abilities and potential through programme management, nurture their growth through increased responsibility and roles, complement capabilities through team work and nurturing, and a value-based and inclusive working environment. Learning through hands-on experience, acquiring knowledge and skills, as well as reflection, are the tools for becoming effective leaders and managers. We have consciously cultivated and cherished diversity as a core value to enhance growth, productivity, collaboration and inclusion.

As we worked towards clearing away the cobwebs in the processes that we were creating for Integrated Slum Development, it became very clear that partnerships and collaborations would be the key to the success of ISD. We would have to learn to apply the principles of mutually beneficial partnerships. Over the years, we have created and nurtured longstanding partnerships with communities, government agencies, academic institutions, civil society, donors, and corporates.

In the next chapter, we'll see how a small and intelligent scheme went onto becoming a path-breaking initiative.

CHAPTER 3

THE REVOLUTIONARY INITIATIVE

In 1973, the Sabarmati river flooded, displacing 12,000 people. The people, who lived along the riverbank, moved inland towards Vasna in Ahmedabad. Its proximity to central Ahmedabad, the city bus terminus of Vasna, and the affordable "grey market" land made Vasna an enticing focus for settlement, and soon evolved into the Pravinnagar-Guptanagar (PG) slum.

In 1991, Saath began engaging with the PG communities – a cluster of four slums with about 1,200 households per slum; a total of 5,000 households. PG was representative of the many growing slums, which were sprouting in and around Ahmedabad and other cities in India due to rapid urbanization. However, PG was unlike SMC (outlined in the Chapter 1), which was, in a sense, established, with little scope for growth. The other reasons were, an absence of basic infrastructure and services, and the community's readiness to work with Saath.

Saath's socio-economic and health surveys conducted in 1991 showed that:

- Residents were recent migrants with a continuing inflow
- It was a cosmopolitan slum with residents from Ahmedabad, Gujarat, as well as from the neighbouring states of Rajasthan, UP, and Maharashtra
- The rate of children not going to school, as well as school drop outs, was high
- The level of basic services was poor
- The incidence of water borne diseases was high
- Immunisation levels were low

- Infant mortality amongst girls was high
- Malnourishment levels were high
- Facilities for pre and postnatal care were lacking.

These findings indicated that there was a dire need for ISD. Apart from the inadequate number of public toilets and water outlets, some residents of PG had come together and installed rudimentary drainage and water lines. Some constructed toilets in their houses with soak pits. This indicated that residents were willing to pay and invest in basic services. There was a glimmer of hope for creating basic infrastructure at the slum level, with investments from within.

Saath had heard of basic services being provided at the household level in a slum in Indore. In 1995, we met Mr. Himanshu Parikh, who had been closely associated with the Indore project. Mr. Parikh had initiated discussions with the Ahmedabad Municipal Corporation (AMC) for a similar project to provide basic services in the slums of Ahmedabad. He was looking for an NGO partner that had worked with and had close ties with slum communities. We were unsure whether it was technically possible to lay water and drainage infrastructure in the crowded and narrow roads in these slums. Mr. Parikh demonstrated how he had overcome this challenge in Indore. We agreed to participate in the design and implementation of what had popularly became known as the Slum Networking Project (SNP) of Ahmedabad.

We consulted the communities in PG on their willingness to pay for household toilets, drainage, and water connections, common paved roads, and street lights. They were keen on getting these facilities and were willing to pay a reasonable price provided their continuous apprehension of being evicted was addressed.

The grassroots project that changed lives

Bhikhabhai and Laxmiben Sondarva moved to Guptanagar in 1977. Forty years ago, they had migrated to Ahmedabad from Amreli district in Gujarat for better livelihood prospects. Bhikahabhai's father and elder brothers had settled in Ahmedabad earlier.

Bhikhabhai had studied up to the 2nd grade and learnt the plumbing and fitting trade as an apprentice with a contractor. He subsequently found employment with the Ahmedabad Municipal Corporation (AMC). Laxmiben has worked as a domestic maid. They had two sons and a daughter.

Before settling in Guptanagar, Bhikhabhai lived in the adjoining Juhapura area. Looking for better accommodation, he met Hirabhai Patel, a landlord in Vasna village, who offered him a 60 square yard plot in Guptanagar for a monthly rent of Rs 12.

"When we came to Guptanagar in 1977, due to paucity of funds, we built a kaccha shack," Bhikhabhai said. "There was no water supply, drainage or toilets. I first built a toilet with a soak pit for the convenience of my wife and daughter. With financial contributions from some neighbours we built a rudimentary drainage line."

"Then, we heard about SNP from Devuben, our neighbour, who was working as a teacher in Saath. She informed us that each household would have to contribute Rs 2,100 as a one-time fee."

Bhikhabhai then enquired about the SNP at the AMC, where he was working, and was told that it was a genuine project in which Saath was a partner. Further, residents would get a 10-year no-eviction guarantee, which could be renewed.

"This increased my trust and I started convincing my neighbours to participate and contribute their fees."

He helped form the Residents Association (RA) and became its secretary. "I knew that the resident's contribution was not going to Saath or the AMC, it was deposited in the RA's bank account. This increased our confidence," he remembers.

Laxmiben became a member of Sakhi Mahila Mandal, spreading the message about the benefits of contributing towards the SNP.

When the physical work started, Bhikahbhai ensured that quality work of laying the water and drainage lines was done. "I knew what was to be done, as I was a plumber and had done similar work for 35 years. When the work was completed to our satisfaction, the RA transferred the contribution to the AMC," he said.

The neighbourhood changed dramatically after the basic service infrastructure was installed. "We upgraded our house gradually. Initially we spent about 3 lakh rupees to build pucca walls, a concrete slab roofing, flooring, a kitchen, and a bathroom with fittings. Last year we added another floor with two rooms at a cost of Rs 6 lakhs. We took a loan from AMC, where I was working, and dipped into our savings to finance the construction," they said, adding with pride, "The value of this house in now almost 40 lakh rupees." They also contributed towards constructing a community temple.

Bhikhabhai and Laxmiben feel that the community has changed too. People from different castes and community interact more. Untouchability has almost disappeared. "People from the Bharwad and Rabari community eat prasad from our temple," they said. "It is also very safe here, especially for women and girls. They have no desire to move to a formal housing complex."

They regret that their Muslim neighbours had to leave Guptanagar after the 2002 riots. "Outsiders came and started burning the houses of the Muslims and they fled," they recalled. "We miss them and we reminiscence about the good times we had at Guptanagar whenever we meet."

How does SNP create change?

SNP was designed as a partnership between the Ahmedabad Municipal Corporation, slum communities, NGOs, and CSR from the corporate sector. Mr. Keshav Verma, the then Municipal Commissioner of Ahmedabad and, Mr. P U Asnani, the Deputy Commissioner, have to be applauded for their vision and support of the SNP. The role of AMC was to design and lay the infrastructure, communities would partly pay for the infrastructure and ensure quality work, NGOs would be the interface between the AMC and the communities, and the corporates could contribute through CSR.

The apprehensions of the community, on eviction, was put to rest with the AMC giving a guarantee of no eviction for ten years, which could be renewed. We pitched for a mechanism by which residents could eventually become owners of the land they occupied. Unfortunately, our request was far ahead of the current thinking. We accepted the ten-year guarantee as the best option. In 2017, the Gujarat government passed a legislation that allows slum residents to purchase the land that they occupied under the Urban Land Ceiling Act.

SNP had two components, creating physical infrastructure and, community development. It was estimated that it would cost Rs. 6,000 per household to create internal infrastructure in slums. Facilitating community participation was estimated at Rs. 1,000 per household. The partners would share the costs as follows:

Component	AMC	Community	NGO	Corporate CSR
Physical Infrastructure	Rs 2,000	Rs 2,000	-	Rs 2,000
Community Development	Rs 700	-	Rs 300	-

- AMC would independently bear the cost of creating external infrastructure.
- AMC dovetailed it's 90:10 toilet construction scheme into SNP
- Saath spent much more for its ISD community development initiatives
- Sharda Trust, the CSR arm of Arvind Mills would contribute Rs 2,000 per household for the first two SNP projects in Sanjaynagar and Guptanagar.
- After the first pilot project in Sanjaynagar, AMC paid Rs 4,000 for physical infrastructure when corporate CSR was not forthcoming.
- It was compulsory for all households to participate.
- Participant households would become municipal tax payers

The communities were willing to contribute their share. In fact, they felt that SNP was a good bargain, as they would benefit much more in comparison to their contribution.

SNP was a path-breaking initiative.

It was structured as a partnership between stakeholders. It was not welfare driven. Slum residents were considered customers who paid for services and therefore brought in accountability in terms of quality and timely delivery. There were naysayers too. Some NGOs felt that the poor should not pay, it was the responsibility of the state to provide services. We were accused of being agents of the World Bank and were promoting privatization of basic services. The local corporators were unhappy, as they felt their power of patronage was threatened.

Our role as an interface between communities and the AMC was sensitive and crucial. We had to be a trustworthy and credible intermediary between AMC and the communities. This involved obtaining approval from the residents for the detailed infrastructure plan prepared by the AMC, ensure quality and timeliness of work, regular payment by the residents and, formation of Resident Associations. Our skills in being credible partners were tested and honed during this period. We simultaneously carried out the health, education, and income generation programmes in the community with the local staff members and, earned their trust and confidence. Similarly, we had a credible record with AMC during our work in building toilets in SMC.

To ensure accountability and avoid corruption, the residents did not pay the AMC directly, but deposited their contribution in bank accounts of Resident Associations (RAs), which had been formed to ensure quality and timely payment. The residents, based on predetermined parameters, made payment to the RAs and AMC. This arrangement gave residents leverage in case the RA, Contractor, or NGO, did not deliver in terms of quality and time allotted. Residents and RAs monitored progress and quality of physical works.

SEWA Bank agreed to provide loans to residents towards their contribution to AMC. However, this proved to be unnecessary, as the average contribution of Rs. 100 per month was an affordable amount. We formed a Sakhi Savings and Credit Cooperative Society to facilitate savings, loans, and residents' contributions to the RAs. Each household became a member of Sakhi and deposited their contribution, which was later to be deposited with the RAs, as per predetermined schedules.

The schedule for payment by the residents and RAs was based on the progress of the infrastructure works. Residents paid an initial deposit of Rs. 500, after the AMC's standing committee approved the scheme. The next instalment of Rs. 500 was to be paid after the drainage lines were laid, and then Rs. 500 after the water lines were laid. The final Rs. 500 was to be paid after completion of work. The infrastructure work was scheduled to be completed within 2 years.

Our strategy for implementing SNP in Sanjaynagar and Gupatanagar was fourfold. First, the design had to be approved by the community. Second, a RA representing all households had to be formed. Third, residents had to deposit their contribution with the RA and subsequently from the RA to AMC. Fourth, residents had to supervise the infrastructure work and ensure timely execution and quality standards. We created a dedicated team for SNP. We also developed a Management Information System (MIS) system in which contributions were mapped with individual households. We did not have any marketing collateral other than verbal commitment.

On the ground, it was an exciting and trying time for our team. Although a majority of residents had in principle agreed to contribute, convincing them individually to part with their hard-earned money was difficult. This was mainly due to their distrust of the AMC. After all, up until now, they had been threatened with eviction by the AMC. Local political functionaries, whose incomes were linked with their powers of patronage, spread false rumours. The three-tiered payment played a big role in convincing the residents to contribute.

AMC decided that Sanjaynagar would be the pilot project for SNP. Sanjaynagar was a smaller slum with 180 households, where a majority of the residents were from the Devipujak community and many former workers of Arvind Mills. Sharda Trust, which was contributing Rs. 2,000 as CSR contribution undertook the infrastructure work independently and not through AMC contractors, which was the standard practise. As the intermediary NGO, we started facilitating community contributions and implementing preventive health and pre-school education programmes. During implementation, AMC and Sharda Trust had differences of opinion, which ultimately led to the withdrawal of Arvind Mills from SNP. However, SNP was completed successfully in Sanjaynagar by 1998.

In Guptanagar, the single RA idea did not work. The seven distinct communities, based on caste and geographic origin, rarely communicated with each other. Internal distrust and power dynamics reigned. For smoother and effective coordination RAs were formed community-wise. Design approval was challenging – objections were raised at every turn – positioning of manholes, placement of toilets, to name a few. Opening accounts for about 1,000 households with Sakhi was a major task. Timely contributions were a challenge. The hundred or so households who contributed initially, set an example for others to follow. Special conditions and alternative arrangements were made for households genuinely unable to pay.

By June 1996, the residents contributed close to Rs. 5 lakhs and were eagerly waiting for the AMC to start the infrastructure work. Then there was a major setback. Arvind Mills withdrew from SNP due to major disagreements regarding its implementation; which meant that their CSR contribution was withdrawn. This was a big blow for the residents of Guptanagar, as well as for us; our reputation was at stake. We decided that we would look for an alternative to the Arvind Mills contribution.

The period from January 1997 to March 1998 was a very trying time for Saath in Guptanagar. Residents were clamouring for infrastructure work to start, and rightly so, they had already contributed their share. The naysayers and critics in Guptanagar started demanding a refund of their contributions. Our patient convincing the community that their monies were safe, and that SNP would happen bore fruit.

Fortunately, our health, education, and income-generation programmes were going on full pace and the community was reaping its benefits. What really helped build confidence in us during those trying times, were the local staff members, mainly women. They understood our commitment to Guptanagar and SNP and convinced the community their contribution was safe. It was deposited in the Residents Association bank accounts, which they managed. They were constantly questioned by their fellow neighbours and were able to convince them to repose their trust in Saath. We lobbied with sympathetic bureaucrats and corporators in the AMC to find a way out.

Finally, in June 1998, AMC decided that it would contribute the additional CSR component per household. This was understandably a big relief for the families and us. Mutual patience and perseverance paid off. Our learning was that public institutions, accountable to its citizens for providing services, such as AMC, did not withdraw from SNP. Neither did Saath, as a NGO, we were accountable to the residents of Guptanagar. Sustaining a partnership between such varied stakeholders, AMC – a bureaucracy, Saath – an NGO, and Arvind Mills – a corporate, required sensitivity, patience, and an acceptance of the different styles of functioning of each partner. "Appreciate in public and criticize in private," has been our mantra to sustain partnerships.

In October 1998 infrastructure work began in right earnest in Guptanagar. There was enthusiasm among the residents and the RAs became active. Then, the next big challenge arose, cutting the boundaries of a few houses to accommodate the drainage and water lines, as well as pucca roads. In the design and plans approved by the communities and AMC, house cutting had been minimized, some houses though, were affected. The RAs and local Saath staff members used all their goodwill and influence to convince the few affected households. The AMC senior engineers were surprised by the residents' readiness to comply with house cutting. When the Rabari community households agreed to do so without any type of compensation, one AMC engineer remarked, "this is history in the making!" Rabaris had acquired a reputation for encroaching on land and never withdrawing without compensation.

Quite a few residents worked as skilled labour in the construction industry. They kept an eagle eye on quality. This caused discomfiture to the AMC contractors, but they could not get away with shoddy work. Recalcitrant residents could no longer find excuses and began contributing. The resident menfolk started assisting the AMC contractors. A perceptible shift in ownership of SNP happened. Residents started viewing SNP as their project, not AMC's or Saath's. The shift from cynicism to trust happened. This was reflected in the increase in participation in all aspects of SNP – financial contribution, convincing neighbours and facilitating the physical work. Guptanagar and Sanjaynagar have become an active part of Ahmedabad city; its residents see themselves as true citizens and not alienated slum residents any more.

Slowly but surely, Guptanagar was physically, socially, and economically transformed. Toilets, water supply, drainage at the household level, and paved roads have led to a significant decrease in water-borne diseases. Overall incomes have increased (data in the survey). Household expenditure on health reduced, and expenditure on education increased. Women, who earlier had to stand in a queue for three hours to fetch water, have more time for productive work. Diffident communities talk to each other. More children go to school, especially girls. Dropout rates have decreased. Residents have started upgrading their houses. More families have bought scooters, fridges, and TVs. Cooking gas, rather than wood, has become the fuel of choice.

How a slum dramatically benefited from ISD

In our study to measure the impact of Integrated Slum Development in Guptanagar, Saath conducted two socio-economic surveys. The first was conducted in May 1997 before slum networking started in which 101 households were surveyed (every tenth household) A similar survey was conducted in December 2000 in which 84 households out of the original 101 of May 1997 with the same residents and families. (The remaining 17 houses were either empty or the residents had changed). For the study, the information collected in May 1997 is considered the baseline data. The December 2000 survey is observed as one indicating the achievements of the ISD programme in Guptanagar. The results were very encouraging. A synopsis of the changes that happened are given in the following tables.

Quality of Life Parameters – Health and Education	Percent (%)
Increase in complete immunization of girls	57.41
Increase in Complete Immunization of children	25.71
Increase of women going for gynaecological check up	11.30

Contd...

Quality of Life Parameters – Health and Education	Percent (%)
Decrease in Home Deliveries	39.95
Increase in Hospital Deliveries	17.89
Increase in children going to school	30.69
Decrease in dropout children from schools	73.68
Increase of adult literates	11.25
Increase in children going to 1–7 Std.	34.78
Increase in children going to 8–12 Std.	38.46
Increase in children going to College	25.00

Quality of Life Parameters – Income, Expenditure and Occupations	Percent (%)
Decrease of Households with incomes of less than Rs 1000 per month	87.50
Decrease of Households with incomes of between Rs 1001 – 2000 per month	60.71
Increase of Households with incomes of between Rs 2001 – 3000 per month	46.67
Increase of Households with incomes of between Rs 3001 – 4000 per month	14.29
Increase of Households with incomes of more than Rs 4000 per month	200.00
Average increase of Household Incomes	66.94
Increase of monthly household expenditure on food	30.83
Increase of monthly household expenditure on conveyance	53.14
Increase of monthly household expenditure on Pan/Bidi/Tea/Entertainment	43.19
Decrease of monthly household expenditure on Rent	30.42
Increase of monthly household expenditure on Electricity	27.34
Increase of monthly household expenditure on Fuel	122.81
Increase of monthly household expenditure on Education	43.49
Increase of monthly household expenditure on Medicine	8.42
Average increase of monthly household expenditure	36.22
Increase in government jobs	27.27
Decrease in private jobs	4.55
Increase in skilled workers	15.09
Decrease in unskilled/casual workers	44.83

Quality of Life Parameters – Income, Expenditure and Occupations	Percent (%)
Increase in vendors	38.10
Increase in recycling/sweeper occupations	83.33

Quality of Life Parameters – Household Assets and Electricity	Percent (%)
Increase in households having a fan	29.03
Increase in households having a TV	38.46
Increase in households having a fridge	150.00
Increase in households having a bicycle	28.89
Increase in households having a scooter	185.71
Increase in households having a sewing machine	125.00
Decrease in households having unmetered electricity connections	38.24
Increase in households having metered electricity connections	65.79

Quality of Life Parameters – House Upgradation	Percent (%)
Households that actually upgraded houses	38.10
Households that planned to upgrade houses	33.33
Decrease in households with one room	37.04
Decrease in households with two rooms	4.65
Increase in households with three rooms	88.89
Increase in households with more than three rooms	80.00
Households that raised house plinth	32.14
Households that added room (s)	32.14
Households that redid roof	25.00
Households that added one storey	17.86
Households that redid floor/walls	39.29
Households that redid/added fittings	17.86
Households that spent Rs 10,001 to Rs 25,000 for house upgradation	21.43
Households that spent Rs 25,001 to Rs 50,000 for house upgradation	17.86
Households that spent more than Rs 50,000 for house upgradation	28.57

Contd...

Quality of Life Parameters – Effect of regular water supply	Percent (%)
Women stating that they saved 1 hour due to regular water supply	30.00
Women stating that they saved 1 to 2 hours due to regular water supply	37.14
Women stating that they saved 2 to 3 hours due to regular water supply	24.29

Replicating SNP

After the successful completion of SNP in Sanjaynagar and Guptanagar, AMC decided to replicate SNP across Ahmedabad city. Standard Operating Procedures for identifying eligible slums were finalized. It was possible for slum communities and NGOs to avail of the scheme. The basic criteria of contributions, partnerships, and implementation remained the same. AMC approached the Indian Institute of Management (IIM), Ahmedabad for suggestions of parameters for selecting NGOs that could take up SNP work.

There was one bottleneck though.

After Sanjaynagar, the AMC appointed its approved contractors to carry out the physical works. However, contractors were hesitant to take up these contracts, as they could not compromise on quality. The residents monitored the works closely and there were no kickbacks. Contractors who took up SNP work did not give it any priority. As a result, the work was slow and did not adhere to the timelines set. Eventually, the AMC agreed to allow outside contractors and the progress of SNP in various slums increased. By 2006, 47 slums with 9,348 households were upgraded through SNP.

The SNP model was highly acclaimed. It won the Dubai International Award in 2005 as a best practice to improve the living environment. There were plans to upgrade the SNP cell to a department level by the Ahmedabad Municipal Corporation. Various international development agencies were keen on funding this initiative. Mr P U Asnani, the Deputy Commissioner of AMC, who had been involved with SNP since its planning stage, was extensively lobbying for institutionalizing SNP within AMC. Alas, this did not happen.

Governance of Municipal Corporations in India is headed by a Municipal Commissioner, an appointed bureaucrat with considerable powers. His or her preferences can significantly influence innovative projects like the SNP, which were not part of the usual municipal services. We have seen the fortunes of AMC rise and fall with the change of Municipal Commissioners in Ahmedabad. At that critical period when SNP was to be institutionalized, there was a change of Municipal Commissioner. The new Municipal Commissioner shot down the proposal.

In December 2005, the Government of India introduced the Jawaharlal Nehru National Urban Renewal Mission (JNNURM) as a major vehicle for urban investment and renewal. JNURM had a component called Basic Services for Urban Poor (BSUP) and Integrated Housing and Slum Development Programme (IHSDP) for providing basic services and housing for the urban poor. It could have been the ideal platform for replicating SNP. However, AMC and many other urban local bodies interpreted the scheme as an investment for new housing. We wondered why? Was it because SNP work is perceived to be tedious and does not involve kickbacks, because it empowered the urban poor with shelter rights, or, was it because building new houses entailed purchase of raw materials such as, cement and steel with associated corrupt practices? We will never know for sure. But JNNURM did slow down SNP projects throughout the country.

It is only now, in 2019, that the importance of incremental housing, which is hugely dependent on the provision of basic services, is being recognized. We are hopeful informal settlements will be included.

CHAPTER 4

A Roof above their Heads

Thirty-seven-year old Chanda Nandkishore Puri is small built and looks fragile. But underneath her deceptive facade lies a strong personality pulsating with enthusiasm and courage. She lives in Ahmedabad with her sister's family. Her husband works as a tailor in a factory. She has three children – two girls and a boy, all going to school.

To supplement the family income, Chanda started rolling incense sticks. But soon disaster struck. She fell sick with chikungunya and could not contribute toward the upkeep of the house they were sharing. Her brother-in-law threw Chandaben's family out of the house. She fought for her share of the house and managed to get Rs 30,000 from her brother in law, which she invested in Sahara Finance. With great difficulty, Chanda's family found a small rented room in the city. But in effect, she was left to fend for herself.

This is a very common scenario that befalls thousands of people in India. So what can they do to pull themselves out of the vicious cycle of debt, poverty and back-breaking labour? Saath came up with innovative ground-breaking methods to tackle this beast. In this chapter, I'm going to show you that it's possible to empower people and give them a roof above their heads at throwaway prices.

But first, we must look at the confusing numbers

A report submitted by a technical committee to the Ministry of Housing and Urban Poverty Alleviation (MHUPA), states that India's urban housing shortage is estimated at nearly 18.78 million households, in 2012. The Statistical Compendium of Slums in India, 2015 prepared by National Buildings Organization, Ministry of Housing and Urban Poverty Alleviation, Government of India, states that 1.375 crore houses in slums have been enumerated in 2,543 reporting towns. Taken together, these reports indicate that 1.878 crore houses have to be built, but also saying that the existing stock of 1.375 crore houses have to be discarded as they are in a dilapidated condition.

Let us assume that out of the 1.375 crore existing housing stock in slums, 3.75 million houses are located on hazardous land and on land which is required for critical infrastructure in cities. That leaves a stock of 10 million houses which can be upgraded and will require an investment of 2 lakh crore (@ an average Rs 2 lakhs per house). Building new houses (@ Rs 7 lakhs per house) will require an investment of at least 7 lakh crores. It does not make any financial sense at all. Why invest/waste an additional 5 lakh crore unnecessarily?

The disintegration of a social fabric

Apart from the financial waste, destroying slums and relocating the residents in distant formal housing flats is socially and economically extremely disrupting. Cities are essentially networks of mutually beneficial social and economic needs that have been cultivated and fostered over the years. Let us consider a few examples. Housemaids, vegetable vendors, plumbers, electricians, gardeners, and rickshaw drivers live in slums near your residence. They charge affordable rates because they do not have to travel long distances to service your requirements. If these people are relocated to distant affordable housing sites, they may not be able to provide services, and if they do, their rates will go up considerably.

Residents in slums build strong social networks and neighbourhoods, between themselves, which ensure that criminal activities are discouraged. By separating the residents and unravelling neighbourhoods, as is usually done during relocation, social networks are disrupted. (Ahmedabad Municipal Corporation uses a lottery method to allocate houses for relocation. Effectively, neighbours are separated and relocated at different sites).

Studies (https://www.ncjrs.gov/App/Publications/abstract.aspx?ID= 145329

https://www.theguardian.com/cities/2016/jan/04/crime-community-designer-social-housing-winnipeg) have shown that the new social housing sites in the USA and Europe have become hubs of criminal activities because they have no social cohesion.

When forcibly moved to a formal house in a location that is socially or economically unsuitable, a slum resident will rent out his/her house, and move to a suitable location, which in all likelihood, will be another slum. While this practice is criticized by the authorities, there is merit in the actions of the affected slum resident. A new house located far away will not compensate for the loss of livelihood and social interaction.

The housing continuum

Migrants drive India's rapid urbanization from rural areas. Migrants, who come to cities seeking better employment, normally have no support system when they come to the city. They migrate based on traditional social networks. Once they have come to the city, they use these networks to find employment, housing, etc. However, the housing that they find through these networks may not allow them to have access to basic services and many of them may end up living on the roadside, or in other areas. A lot of the energy that migrants come to the city with gets spent on accessing basic services rather than useful economic activities.

The following diagram shows the way in which migrants meet their housing needs as they transition to become long term residents in a city.

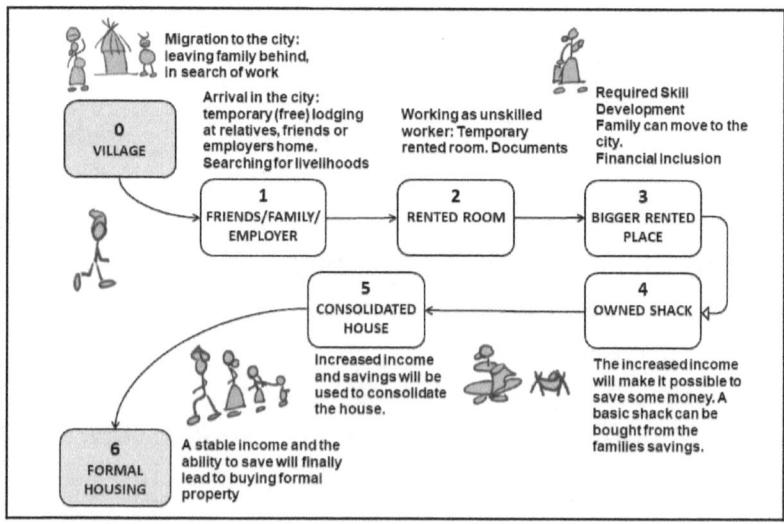

As a migrant settles in a city, the need for housing changes based on personal, livelihood and social needs and choices. The various stages during which a migrant becomes a resident are described below.

Stage 1 – Arrival in the City: The migrant finds temporary (free) lodging with relatives, clansmen, friends or other contacts in the city for a period of one to two weeks. He then looks for work, which will in all probability be in the informal sector. At this stage, daily expenses are low and the only area of concern is travel expenses and food. There is no expenditure on housing.

Stage 2 – Shared Rented Room: After finding work, the migrant shifts to a shared rented room in a slum, where the rent will be about Rs 500 per month. General duration of stay is around 6 months. Monthly income at this stage is stable, meaning that the migrant has paid work for at least 20 days a month.

Stage 3 – Individual Rented Place: Once the migrant's employment and income are stabilized, his family moves into the city. He then rents an individual house in a slum. With more family members being able to work, the household income goes up to Rs. 10,000 per month. A private single room will cost Rs. 1,000 to 1,500 a month and a two-room house will cost around Rs. 3,000. Duration of stay is from one to five years.

Stage 4 – Owned Shack: The income of more than Rs. 10,000 per month makes it possible to save some money. After a while, the migrant, now better acquainted with the city, purchases a basic shack in a slum. A kaccha (non-concrete) house costs between Rupees 1,50,000 to 2,00,000, depending on the area. A family lives there for approximately two years before they start to incrementally upgrade their house.

Stage 5 – Consolidated House: Increased income of Rs. 10,000–12,000 per month and savings are utilized to consolidate the house. In a period of 5 to 7 years the family upgrades their house to a semi-pucca or pucca house in 1 or 2 renovations. The cost for the renovation is between Rs. 1 to 1.5 lakh in total or Rs. 25,000 to 30,000 per year.

Stage 6 – Formal Housing: A stable income of more than Rs. 15,000 per month and the ability to save finally leads to buying formal property. The family then sells their consolidated house, which by now is worth around Rs. 4 lakhs. A 1RK apartment in a legal housing scheme for Economically Weaker Sections (EWS) and Low Income Group (LIG) families, costs between Rs. 4 to 6 lakhs.

In the housing continuum, houses in slums are the only affordable option for a migrant. If slums are destroyed, where will the migrants be housed? Isn't it smart to upgrade houses in slums rather than discard them?

How Saath facilitated both, incremental housing and new housing

Incremental housing

The phases in the housing continuum show that slums provide the bulk of affordable housing stock. We have been advocating for the upgradation of houses in slums and, by doing so, converting slums into respectable living spaces.

Effect of Slum Networking on Incremental Housing

In the Guptanagar project, the transformation of the slum proves the point.

After being assured by the authorities that they would not be evicted, they not only contributed, but supervised the work for the basic services of water, drainage, electricity, sanitation and paved roads. With the amenities in place, the residents started incrementally upgrading their houses in a big way. Floors were tiled, rooms and floors were added, roofs were cemented and bathing areas constructed. Kuchha dilapidated houses started becoming pucca houses. But this did not happen overnight. Residents upgraded their houses at a pace convenient to them. Upgradation happened when money was available, when children were not going to school and, alternative accommodation could be arranged.

Each household was upgraded differently as per the needs of the family. Neighbours negotiated to ensure that the integrity of common spaces was maintained. Home owners identified competent local contractors, got cost estimates and supervised the construction. Finance for upgrading was obtained through various sources. Savings were used, loans taken from family, friends, and employers, and eligible members from the Sakhi Mahila Credit and Savings Cooperative. The local economy received a boost due to the increase in construction activity. Even today, almost 6 years after the services were provided, some houses are still being upgraded. Our socio – economic survey of 2000 indicates that an average of Rs 20,000 was spent by each household for upgradation. Over the years, residents have made more investments. The financial values of the houses also increased since they were part of AMC's Slum Networking Project. The Residents Associations took up waste collection contracts. Today, Gupatanagar, like other slums which participated in SNP, no longer looks like a slum.

Incremental housing upgradation did not disrupt the livelihoods and social networks of the residents of Guptanagar but enhanced them. The residents felt more secure and this is reflected in the socio-economic survey of 2000.

The fear of eviction, though greatly reduced through the 10-year no-eviction guarantee given by the AMC, still persisted. In 2017, the Gujarat Government legislated that owners could purchase the land on which their houses were located (land surrendered under the Urban Land Ceiling Act). This is a very progressive step and will increase incremental housing significantly. Residents of Guptanagar are now legal owners of their houses.

Making incremental housing work does not constitute rocket science. It requires a form of security like, a no-eviction guarantee, patta transfer or a mechanism to purchase the house or land parcel. The next step is to provide the basic services of water, sanitation, paved roads, and electricity supply, which is paid for by the residents themselves. The third requirement is a mechanism for financing the cost of physically upgrading the houses. Fortunately, the National Housing Bank has allowed Housing Finance Companies to lend to the informal sector, for purchase of new houses, as well as for upgradation.

Incremental housing does not require a developer. Permissions required for new construction, purchase of housing materials, and a minimum period of three years is what it takes. Yet, our authorities do not encourage incremental housing.

Providing new, affordable housing

2009 was a watershed year for housing finance in India. The regulator for housing finance, the National Housing Bank, made a policy change through which housing finance could be made available to people in the informal sector. For the first time, people who did not file IT returns or have salary slips, could get housing loans. This meant that rickshaw drivers, vegetable vendors, construction workers, small shop owners, drivers, and all those who were not employed in the formal sector, could avail of a housing loan. The affordable housing sector had just opened up.

A study by Monitor Inclusive Markets in June 2012 indicated that the real demand for housing was in the affordable housing segment. High Income Group (HIG) housing was not a growing sector. The opportunity for real estate developers, was in the affordable housing sector.

In Ahmedabad, Mr Sanjay Shah, a real estate developer formed DBS Affordable Housing Pvt Ltd in 2009, a company that would focus on building affordable houses only. This was a trend-setting initiative. The difficulty was that DBS did not understand their potential affordable housing customers. Developers had dealt with customers for MIG and HIG housing and understood their housing aspirations and, designed, priced, and built houses accordingly. The need was to get a grip on the housing designs and price points that would match the aspirations of affordable housing customers, many of whom would be from the informal sector. The second challenge was to market these houses. Developers did not understand the characteristics of the informal sector, as a housing market. Developers had to design an affordable housing model, which would work for the informal sector that constitutes the LIG segment.

We realized that the housing needs of the informal sector we were working with, could not be met by the housing stock available in slums and chawls. At the same time, families in slums were aspiring to buy legal affordable houses. When DBS approached Saath to jointly design a business model for affordable housing, we had understood the potential demand and the need.

We then conducted a snapshot study in December 2009 with 100 families living in informal settlements to understand how the Base of the Pyramid segment used their housing and living spaces, their aspirations for a new house, and the difficulties they perceived in purchasing a new house.

The key learnings from the snapshot study were that potential customers could pay an installment of around Rs 2,000 per month, purchase a house costing between 2 – 4 lakhs, would be willing to relocate beyond 5 km, wanted a house with one to three rooms with a separate kitchen, bathroom, and toilet. The study suggested that potential customers would require assistance to enhance their incomes, access housing loans, support for acquiring basic KYC documents, and opening bank accounts.

Based on the snapshot study, we designed a business strategy for affordable housing. The model had three components. The first was design and construction of affordable housing, the second was facilitating access to housing finance and, the third was community engagement to enhance incomes and improve creditworthiness of potential home buyers.

In January 2010, DBS organized a community housing design workshop with experts from the real estate sector, housing finance, architects, and civil society. The objectives of the workshop were to validate the DBS business model, to identify potential institutional linkages for technical and financial resources and seek collaborations with stakeholders, which could include government agencies, investors, housing finance companies, foundations, and NGOS.

From the snapshot study and design workshop, we understood that the key component was to create awareness among potential customers about the opportunities for purchasing their own houses, as well as the process for buying a house. We knew that traditional marketing methods would not work, as this was a completely new segment where information and knowledge about affordable housing was in a nascent stage. The segment had to be made aware that there were developers who were building affordable houses, there were housing finance companies that would give loans, and community development initiatives, which could help enhance their incomes.

Griha Pravesh

To create this awareness, we promoted Griha Pravesh (GP) a social enterprise to create awareness about affordable housing. GP facilitates marketing, sale, and purchase of affordable housing, and provides guidance on all aspects related to affordable housing. GP works with developers, housing finance companies, government agencies, and potential customers.

To identify homebuyers and create awareness about affordable housing, GP maps geographical areas where potential customers are located. Then these areas are surveyed to get household information. The information is analyzed to identify families who are interested in buying a house and those who could buy a house. GP contacts these families and through one-on-one meetings and group seminars, informs these families about how they would be able to purchase a house.

A detailed household financial diagnosis shows the income and expenditure pattern, status of savings and assets and, occupational growth. Based on the diagnostic report, a blueprint for purchase of a house is prepared. This involves preferred location, house size, quantum of housing loan required, down payment to be mobilized, KYC documents to be procured and skill enhancement for increasing incomes. The family is introduced to suitable affordable housing projects and site visits are organized. The family is acquainted with technical specifications, housing terminology such as built-up area, super built-up area, carpet area, No Objection Certificate, and maintenance fund. The family is familiarized with a variety of loan offerings from housing finance companies. After the family decides to buy a house, GP provides support for opening a bank account, procuring KYC documents, availing a housing finance loan and enhancing livelihoods.

GP partners with affordable housing developers by providing business development services. These include viability studies for specific sites, designing products, marketing strategies, on-the-ground marketing activities, and customer engagement during launch of projects.

By working with housing finance companies, GP provides customers with different housing loan options. If required, GP facilitates housing loans by procuring KYC documents, follow-up on loan installments and opening bank accounts. With Foundations, GP promotes financial literacy regarding purchase of affordable houses. GP as a social enterprise earns it revenues through fees charged to customers, developers, and housing finance companies.

Chandaben met Griha Pravesh (GP) staff when they visited her house while surveying potential home buyers in the Bapunagar area of Ahmedabad. When they told her, she could purchase her own house, she did not believe them and said that with her household income, she was destined to live in a rented room. The GP staff explained to her that houses she could afford were available and she did qualify for a loan. Chandaben consulted her husband, who was extremely supportive, and they visited a few sites and finally selected a one room house with a kitchen, bathroom and toilet in the Umang Vatva scheme. The cost of the house was Rs 5.65 lakhs. She took a loan of Rs 3.5 lakhs from a Micro Housing Finance Company, with a fifteen-year tenure, and to make up for the balance down-payment, she sold her jewellery, encashed the Sahara investment (which had doubled) and borrowed Rs 1 lakh from her father. She pays Rs 4,200 as an instalment towards the loan from their monthly income of between Rs. 15,000 to Rs 20,000.

When we met Chandaben, she was decorating the entrance of her house for the Navaratri Festival. She borrowed chairs from her neighbour and made sure that we were comfortably seated in the only room which was furnished with a bed, TV, a sewing machine, and a newly bought cupboard. She said that the new house had transformed her family's life. Her social status was enhanced and after many years, her relatives had started visiting her. They were praising her ability to have purchased a house. She had become a role-model for the women in her extended family.

We asked Chandaben whether paying the monthly installments was a challenge. She said that as she could not roll incense sticks in the new house, she had started taking tailoring job-work. As she could store more material in the bigger house, she had started outsourcing stitching to her neighbours as well. Subsequently, the family's monthly income had increased. The new house has raised aspirations. Her son wants to buy a bigger house with 2 rooms for the family when he starts earning.

Chandaben was effusive in her appreciation for the GP staff who had guided her. Apart from finding an affordable house and facilitating the loan, GP staff had helped her get the KYC documents required for financial transactions. Her smile radiated happiness and hope, while bidding us farewell.

Griha Pravesh has operated in Ahmedabad, Vadodara, Kalyan, Jaipur and Varanasi, and reached out to almost 60,000 people. Partnerships with almost 40 developers and 9 Housing Finance Companies have been established by GP. 3,500 people have bought a house through Griha Pravesh.

'Who can imagine a rickshaw driver buying a house?'

A tall man with a big smile, forty-three-year old Devilal migrated to Mumbai from his village in Dungarpur district, Rajasthan, in 1985 to work as a diamond polisher. In 1992, he migrated to Ahmedabad after the diamond-polishing factory closed down. He started a vegetable-vending stall, but arthritis of the legs prevented him from standing for long hours. As he knew how to drive, he decided to ply rickshaws and today owns one.

Devilal lives in a rented house in the Maninagar area of Ahmedabad with his wife, a college going daughter and a son who goes to school.

They pay a rent of Rs 3,500 for a two-room house. This is his fourth rented house. His wife earns Rs 6,500 a month as a domestic maid and Devilal earns about Rs 25,000 a month.

Devilal decided to buy a house after meeting the staff from Griha Pravesh (GP). He said that they had cajoled him into visiting a newly launched affordable housing site in Vatva called Umang Utsav. With a sheepish smile, he said that the free gifts offered to visitors on the launch date were a major attraction. He saw the sample houses and the affordable prices and decided to buy a one room house. He told his relatives about his choice and they also had a look and ultimately five of his relatives decided to buy the larger two room houses. Devilal was also persuaded by his family to buy a two-room house.

Devilal's two room house costs Rs 13 lakhs. He has got a loan from Central Bank for Rs 11.88 lakhs with a fifteen-year tenure. He will pay the balance down payment from his savings and surplus income. He says that the monthly instalment of Rs. 7,500 is manageable from his family's monthly income. He was proud of his credit history, as he had repaid loans that he had earlier taken to buy a motorcycle and a mobile phone.

Devilal said that were it not for GPs encouragement, he would never have bought a house. He had visited a scheme once five years ago, but it was too expensive. Recently he had passed by the site and saw that hardly any construction had happened. He said that GP showed him a genuine scheme, helped him get documents for the housing loan and guided him in the purchasing procedure.

When we asked him how he felt about moving into his own house within a year, Devilal shot us his usual big smile. 'Who can imagine a rickshaw driver buying a house?' Now he is role model for his extended family. He said that he recommends GP to all his passengers!

Housing resource centers

We have established "Housing Resource Centres" (HRCs) in Ahmedabad, Jaipur, and Varanasi, which act as a one-stop centre where people can be linked with affordable housing. The activities at the centres include assisting people with home buying, rental housing, community mobilization, and research on migrants in cities.

To understand migrant communities, HRCs have studied and profiled migrant communities at the city level, to identify where and in what conditions the migrants live, what work they do, and how they make their choices regarding housing in the city. After the profiling, we identified target groups and presently are in the process of planning and organizing a detailed survey on social, economic, and housing issues of these migrants.

We found that migrants live either at worksites, open spaces, or in rental housing. Each of these ways of living offer some benefits, as well as some disadvantages. For example, people who live in the open normally have no expenditure on "housing" as we consider it, such as rent, tax bills, or electricity bills. However, because they are living in the open, they end up spending more on things such as, sanitation, by using pay and use public toilets, etc.

Each migrant makes this housing choice based on his or her own needs and affordability. Therefore, when a housing solution is presented to these migrants, each migrant has their own view on what they require from it. As an example, night shelters are proposed as options for people who live in the open. But the people who live in the open are often living with families, and night shelters have no provision for a family to stay together. Therefore, the family chooses to stay in the open, even though they have the option of a night shelter.

HRCs have two major functions. The first is to assist the people who need housing services by connecting them to these services and, secondly, conducting research into how these services can be improved and their outreach maximized. The housing resource centres are doing this through their partnership with other organizations such as Mod-Roof, a company that is developing modular roofing designs for people who have houses in slum; the Affordable Housing Mission, the nodal agency for implementation of Pradhan Mantri Awas Yojana in Gujarat; and private developers who are building affordable housing schemes. They also take into account the requirements of the people who need housing, keeping their capacity and affordability in mind.

HRCs plan facilitates affordable housing in an incremental and participatory manner for 4,000 households across these three cities. We aim to use the research on housing conditions for migrants to create a platform for advocacy for these vulnerable communities, as well as to promote participatory planning between the government and migrant communities for the resolution of common problems.

Lighting up Lives

Domestic electricity connections were not a part of the SNP package of services. In Ahmedabad, electricity was supplied by a private utility, Ahmedabad Electricity Company (AEC), now Torrent Power. In Guptanagar, as in other slums and chawls in Ahmedabad, most houses had an illegal/informal electricity connection, provided by informal operators who illegally sourced electricity by tapping nearby electricity lines.

We approached AEC and asked them to give electricity connections to residents of Guptanagar. The response was not very encouraging. AEC officials believed that slum residents preferred to steal electricity, rather than pay for it. They did not consider slum residents genuine customers.

We tried to explain that slum residents stole electricity because it was not legally or formally available to them. They did not have the requisite documents to prove legal ownership of residence and, that the connection charges, which ranged between Rs. 5,000 to Rs. 10,000 were not affordable. We explained the mode of payment to the informal suppliers – Rs. 50 per electrical point per month. Normally each slum household would take three points, one each for a fan, a tube light and a TV. They paid Rs. 150 per month for unreliable tapped electricity to the informal supplier. AEC was however, still not convinced that slums were a market for electricity.

Fortunately, in 2001, the USAID Slum Electrification Programme started exploring ways of providing electricity to the urban poor in

Ahmedabad. We joined them in the research and formulation of a proposal.

We proposed a pilot project in which households that had participated in the SNP programme would qualify, since they had proof of residence by way of tax bills. The pilot project would provide 1,000 connections at Rs. 2,000 per connection. USAID would subsidize the balance connection cost to AEC. AEC would install meters with each connection. The pilot project had two objectives, first to check if residents would pay Rs. 2,000 for a connection and, if the monthly bill would be affordable. On both counts, it was a huge success. All the 1,000 connections were taken in a short time and the ensuing monthly bill ranged between Rs. 150 – Rs. 250, which was affordable. **Moreover, an electricity bill is a very important document as proof of ownership; it was now available to them.** We had crossed a major hurdle.

However, AEC was not fully convinced that just Rs. 2,000 was the affordable subsidized cost price. They asked us to do a larger pilot project where they proposed a connection cost of Rs. 5,000.

After the Gujarat riots in 2002, we had started working in the Juhapura area of Ahmedabad. Juhapura, a large Muslim ghetto, had been neglected in terms of infrastructure. There was a large electricity mafia present. We had heard that the AEC officials were afraid of entering Juhapura. We approached the residents with an offer for a formal electricity connection at Rs. 5,000. The answer was a resounding "YES!" We facilitated the electrification of almost 4,000 houses in Juhapura. Sakhi Cooperative provided loans at 18% interest for electricity connections. We neutralized the resistance from the mafia by training them to become contractors, to provide internal electrical wiring, as per AEC standards.

In the meanwhile, another obstacle cropped up. After the Gujarat Earthquake, regulators ruled that only houses with the Building Use (BU) permission from the civic authorities could get an electricity connection. Houses in slums and chawls do not qualify for BU permission. We filed a PIL in the High Court of Gujarat, with a citizens group, requesting exemption of BU permission for houses in slums and chawls to avail of legal electricity connections. The High Court agreed with us; slum electrification proceeded.

After Juhapura, AEC was convinced. On their own they reduced electricity connection charges to Rs. 2,000 payable in instalments. Since the demand for electricity was substantial, almost 85 percent of households in slums and chawls took connections. **Theft reduced considerably, and demand increased.** AEC's revenues increased enormously. It was a classic example of C K Prahlad's model of a market at the Bottom of the Pyramid, where margins are low, and volumes are high.

Once again, through an innovative programme with relevant stakeholders, and strong partner engagement, we were able to demonstrate that slum residents are genuine customers, willing to pay for affordable and quality services. They do not want welfare doles.

Chapter 5

Livelihoods – Teaching Communities How to Fish

Training programmes that change lives

"One of the reasons I took up a beauty course was because I would have better working hours," said Aagnya Vyas.

Aagnya Vyas lives in Ghodasar with her parents and a brother. Her mother sews for a garment outlet while her father is a maharaj at the Lambha temple. In class 8, she too joined her mother at the garment outlet. After graduating from school, she worked at Dominos for almost five years. The long hours, especially during festival time, left her exhausted. She often got home only by 11 pm.

When representatives from Udaan gave her mother a pamphlet with details of a training programme, Aagnya quickly decided to take up a course in beauty parlour training. "The fees could be paid in installments. We could afford to pay Rs. 250."

On completion of the course, she was placed in a beauty parlour where she worked for a few months. During this period, she got hands on experience in all the beauty treatments.

"I was, however, not keen to just work in a beauty parlour," she said. "I wanted to do other things. I spoke to one of my trainers, Bijal Ma'am and she informed me about an opening in the Maruti Service Department. She told me attend the interview and take it from there. I was offered the job, which I took. My job was to enlighten the customers as to the necessity of getting a coating inside the vehicle. I underwent training. I enjoyed the work and stayed on for a year."

Today, she works at Cadila-Zydus Pharma Company. She underwent training and then was given supervision of the automatic packaging of injections. The timings are also favorable. She starts at 6.30 in the morning and is done by 3.30 pm. Plus, she is picked up and dropped off by the company bus.

"Although I ended up doing something completely different after my training at Saath, it was *because* of that training that I am now confident," she said. "I feel capable of learning and going into new areas."

Saath's work to enhance livelihoods has encompassed employment in the formal and informal sectors, and micro-entrepreneurship. Our interventions have been in training, placements, mentoring, and creating social enterprises. We created partnerships with state governments, private sector, slum communities, civil society, and funders to scale our impact.

Formal sector training

By 2005, the Indian economy was booming, and the service sector was creating employment opportunities in large numbers across senior, middle and entry level positions; but surprisingly, there was a critical shortage of human resources for the entry-level positions. The various graduates that the Indian education system produces were not willing to take up the entry level positions, which they felt were beneath their degrees and qualifications. These entry level positions were across the retail, IT and IT enabled, hospitality, logistics and financial sectors. The profiles in these entry-level positions were for waiters, data entry operators, housekeeping staff, sales persons, etc. These profiles did not require a college degree. A basic 10th or 12th grade qualification was adequate. What was missing were the skills to be functional in these entry-level positions.

We sensed an opportunity to enhance employment opportunities for the numerous youth who lived in low-income neighbourhoods. These youth had dropped out of school in the 10th or 12th grade, mainly due to financial constraints and lack of family support. They were not willing to take up manual labour work. They did not want to become masons, electricians, plumbers or carpenters. They were not willing to join micro-enterprises their parents had nurtured. They could not qualify for formal sector jobs, which required a college degree. They had become unemployable.

We heard about the pioneering work done by Dr. Reddy's Foundation for youth employability in Hyderabad. Dr. Reddy's foundation had designed and piloted a 3-month training program, which would make young people who had dropped out of school, employable for entry-level jobs in the service sector. With the help of the American India Foundation, we decided to replicate this model in Ahmedabad. In 2005, we discussed the model with Mr. Anil Mukim, the then Municipal Commissioner of Ahmedabad. He was convinced and gave his consent for Ahmedabad Municipal Corporation as a willing partner.

We structured this youth employability training program as a Public Private Partnership (PPP) model called, Udaan. Ahmedabad Municipal Corporation provided training centres as they had vacant properties such as, schools and community halls, and provided part funding for the training. We mobilized the youth, conducted the training and ensured placement in various service sector jobs. The American India Foundation supported us with part of the training costs. We had insisted that the youth pay a fee of Rs. 500 towards the training. In 2005, the cost of training was Rs. 4,500 per participant.

We approached CAP Foundation to provide the syllabus, as well as for handholding support, to orient the trainers in pedagogic skills.

A well-furnished and equipped training centre was set up at the Liladhar Bhatt Community Centre in the Behrampura ward of Ahmedabad in 2005 where we started Udaan.

UDAAN – a unique and effective campaign

All Udaan courses were a three-month duration. Participants were trained in functional job requirements, based on market needs. Training was activity based and included classroom learning and practical training for which appropriate equipment and teaching software was provided.

The curriculum was based on requirements specified by HR managers in the market scan. It was dynamic and reflected the changing needs of industry. There were five components:

1. Technical component – training for various sectors
2. Basic English – speaking and understanding
3. Life Skills Module – preparing participants to be comfortable in work situations
4. Basic Computer knowledge
5. Work Readiness Module

The Life Skills Module included positive thinking, self-esteem, communication, team work, decision-making, SWOT analysis, goal setting, gender issues and time-money management; all essential for confidence-building.

A specially designed "Communicative English" module, which included public speaking and inter-personal communication, prepared participants for the communication challenges at the workplace.

Basic understanding of computers was an integral part of the training. Today, job descriptions in all service sectors include being computer savvy.

The "Work Readiness Module" taught the participants how to prepare a resume and face an interview. They were briefed on work culture, coping with workplace related problems and maintaining a balance between personal and professional life.

Why UDAAN worked

Udaan training was completely based on market needs. The market scan was a detailed understanding of HR requirements at the local level through interactions with leaders and HR heads of industry. It indicated both, requirements across sectors and specific skills required for each job. The sectors that were included in the market scan were Hospitality, Customer Relations & Sales Marketing, Multi Skilled Services, IT Enabled Services, White Goods Services, Health Care, Automobiles and Fashion & Accessories.

Interesting road shows articulating the training and job placement initiative, by Udaan, were held in and around slums. Questions regarding the courses were clarified and interested youth enrolled.

Based on their interests, abilities, and education levels, participants were inducted into specific courses and started their two-month classroom training. The training methodology consisted of theoretical and practical learning. Work situations were simulated and emphasis given on peer-to-peer learning.

Faculty members who were middle management, conversant with industry requirements, conducted the training. They mentored the participants through the entire training period. They developed the existing business network for placement of participants.

Guest faculty, normally HR managers from various industries, informed them on latest workplace requirements and interacted with the participants. They also helped in placement of participants.

Towards the end of the second month of training, participants were given assignments in which they visited and interacted with managerial staff in companies to gauge the skills required in their chosen courses. The feedback from managers was incorporated into the training, which improved employability skills.

The "work readiness" module was the stage before final placement of participants. It prepared participants to cope with work place situations. Participants worked for about 10 days in a business setting. Their feedback to the faculty members on additional skills required, based on their experience, was valuable to the course.

The students were finally placed for full-time employment. The place of employment was identified either by the faculty member, by the participant during his assignment, or through a guest faculty member. Later, Udaan attracted employers from established firms.

Counseling was provided to participants after placement to help them cope with their workplaces. Interested participants could enhance their skills through advanced learning modules.

Udaan proved to be a successful program. Between 2005 and 2006, about 400 youth were trained and 300 people got meaningful employment. We felt that the Udaan module could provide effective livelihood options to young people who could not go to college. The PPP model had proved to be workable and created win-win situations for all partners. We were ready to scale up.

Scaling up Udaan

Imtiyaz Qureshi lives in Isanpur. He's studied up to class 8 and after that, worked in a box-making factory, earning Rs. 30 a day (Rs. 900 a month).

In 2008, a friend of his, told Imtiyaz about Saath's 3-month training programmes and subsequent placements called Umeed. He visited the centre where he was counseled and a course on Customer Relations and Sales was suggested to him.

"I leant how to operate a computer, how to type and operate Power Point, Micro Max, Excel," Imtiyaz said. "I also how to adapt my behaviour and attitude when talking and dealing with customers. I used to go about shabbily dressed. I learnt how to dress and behave in a fitting manner. I also learnt basic English, how to say, "How are you? Fine. Thank you, etc."

After his training, Imtiyaz was placed at the Airtel Page Point call centre where he drew a salary of 3,500. This was in 2008. "I worked there for ten months and I was approached by a NGO, Azad Education. I gave lectures on Personality Development. I worked there for six months, after which, I joined Saath and worked as a team leader for mobilization of students for the course I underwent. I visited the 30–35 training centres that were being run. I was also guest lecturer at the course where I shared my experiences and the changes brought about in my life, in the past two years.

"In 2011, I joined Adani Gas Ltd. as a general agent. I am still working for them. My salary was 3,750 when I joined; I now draw Rs. 12,000. I am also working for another NGO, Merapath Education Ltd., a Delhi based organization. We work on digitization in villages under the government's call for "Digital India." I am presently working at both places. I also assist in procuring documents like the PAN Card, Aadhar

Card, etc. At Adani's I am now a senior executive at the emergency help desk. Our work is computerized, we receive complaints of gas leakages, fires, etc., and inform the mobile teams to attend to them. We work shifts. This allows me to work in the villages for the digitization work."

"We give digital competency certificates in the villages. I am now Head of Merapath Education Limited in Gujarat; I have a partner and we work in all the villages under the Pradhan Mantri Grameen Digital Saksherta Abhiyan. I basically overlook the programme; our team does the work."

"I have leant a lot in Umeed. Although I work two places, I do make the time to assist Saath in whatever way I can, wherever invited, even as a guest lecturer. I also carry out campus interviews for vacancies in Adani. Besides Adani, I also assist in placements with third parties with whom we have tie ups. We also help in getting job work for those who can work from home. I salute Saath for all the good work being done and especially for always welcoming those who have moved on after working with them. I am remembered, invited for programmes and lectures. This doesn't happen anywhere else."

In 2007, Saath approached the Urban Development Ministry of the Government of Gujarat and presented the Udaan experience as an effective PPP model for creating livelihood opportunities for unemployed youth across urban Gujarat. Mr. S R Rao, the then Urban Development secretary, approved of the model and asked us to prepare a plan for Gujarat. Mr. Mahesh Singh, the then Deputy Secretary, Mr. I P Gautam, the then Municipal Commissioner of Ahmedabad and Mr. Shanker Venkateswaran, the Country Head of the American India Foundation gave significant inputs in designing the plan for Gujarat.

"Umeed" was the name given for the Gujarat Youth Employability Programme. Like Udaan, it was a PPP model, with the Urban Development Department (UDD), Govt of Gujarat, Urban Local Bodies (ULBs) and NGOs as partners. The UDD was to provide part funding of Rs. 3,000 through the Government of India's Swarna Jayanti Shahari Rojgar Yojana (SJSRY). Foundations were to provide part funding of Rs 1,500 and participants had to pay Rs. 500 as fees. ULBs were to provide training spaces. (Later, in some locations, training spaces were rented. The coordination was to be carried out through the Gujarat Urban Development Company (GUDM).

For implementation of Umeed, we proposed a three-region division in Gujarat, namely, Central and North Gujarat, South Gujarat, and Saurashtra and Kutch. A nodal NGO, with experience in implementing youth employability program, at scale, would manage Umeed in each region. We were selected for Central and North Gujarat, CAP Foundation for South Gujarat and, Aid et Action for Saurashtra and Kutch. Nodal agencies could appoint credible local NGOs as training partners in particular towns.

Umeed became an effective and popular employability program. Between 2007 and 2012, almost 1,90,000 youth were trained and 1,25,000 were placed through 170 training centres across 95 cities of Gujarat. We trained 45,000 youth and placed about 30,000 through 45 training centres in Ahmedabad, Vadodara, Nadiad, Mehsana, Palanpur, Anand, and Kalol. The American India Foundation supported us through part funding for the training.

Our experience of Umeed was bittersweet. The positive aspect was that we had been able to design a program that had significant impact. From working only in Ahmedabad, we were able to successfully scale up to seven towns. We were able to build managerial and implementation capabilities for the programme. We had demonstrated how PPP models could effectively work in the development sector. The negative aspect was the financial shortfall of about Rs. 1 crore that Saath had to bear due to non-payment of Umeed dues by GUDM. By 2009, the procedures and benchmarks in the Umeed program were unilaterally modified due to a change in leadership in UDD and GUDM. The main change was that ULBs were asked to disburse payments. This led to unethical practices, which we could not comply with, which led to non-payment for Umeed services provided by us. Our hard-learned lesson was that while working with the government, a change of bureaucrats can often result in adverse impact on NGOs.

In 2008–2009, we worked with the Rajasthan Mission on Livelihoods and provided similar youth employability training to 9,100 youth in the Barmer, Jalore, Jodhpur, Sirohi and Nagore towns.

In 2009, the Government of India established the National Skills Development Corporation, as a public private partnership to focus on the 30 percent skill gap in India's workforce. Youth employability training became a mainstream national activity. The private sector was invited to participate in skill training. State governments gave skill training a priority with generous funding. We felt that our objective of demonstrating how a successful youth employability-training programme could be scaled up and be part of the mainstream processes, was met. We decided to focus on the large gaps that still existed in skill training in the informal sector.

Informal sector training

Rakesh Patel lives in Ranip, Ahmedabad with his parents and brother. His father works in a mill on a contract basis. After failing his 10th standard exams, Rakesh joined his father at the mill only to leave in a matter of months. He joined the ITI (Industrial Training Institute) for a DTP computer course, but that too didn't work out. All he learnt was how to turn the computer on and off. In 2015, when Rakesh moved to Bapunagar, Ahmedabad. he connected with his uncle, a faculty member for the Plumber's Training course, who told him about Saath's training programmes. He quickly got onboard and started with the electrician's course. This helped him get a number of small wiring jobs.

Subsequently, he heard about the other courses on Computers, Retail Management, Customer Relations and Sales, etc. He soon took a course on Retail Management, and, on completion, was offered a placement.

"I, however, had been watching programmes on television and had learnt about the NGO sector," says Rakesh, "and spoke to my counsellor about my desire to work in the sector."

It was not long before Rakesh was offered a field workers position in the Nirman programme run by Saath.

"My job was to spread awareness about the programme and convince people to join. I worked for six months in the field. Since I had done my basic computer course, I was given an opportunity to work making reports. Due to personal financial problems I had to leave Saath.

But I am currently working in an agency that carries out inspections for insurance companies. This has been possible because of my fieldwork and report writing experience at Saath. I am also studying for my B.Com."

When training for entry level formal sector jobs became mainstream, we decided to concentrate on similar jobs in the informal sector, especially in the construction industry, which employs almost 10 percent of India's workforce. Traditionally, training in the construction industry for masons has been informal, where a master mason teaches apprentices in an unstructured way. The apprentice starts as an unskilled worker and gradually acquires expertise to become a skilled worker. Training can take anywhere between 3–5 years depending on the level of interest and commitment of the master mason and the apprentice. Even after acquiring the expertise, a skilled worker is often at the mercy of a contractor who is the sole provider of work. The skilled worker does not have any type of formal certification and cannot prove her/his skills unless referred by another contractor.

Nirman

We designed a three-month training programme called "Nirman" to improve skills, working efficiency and proficiency of the semi-skilled male and female labourers of the construction sector and help them get better wages and better contracts. Nirman facilitates registration with government departments and enables social security. Nirman upgrades the expertise of semi-skilled workers through theoretical and practical training and most importantly, certifies the training. As a result, the construction worker can get higher wages and has more options for employment, as s/he is not dependent on the good graces of a contractor.

The semi-skilled professions addressed by Nirman are in masonry, plumbing and electrical wiring.

Participants of Nirman are made aware and mobilised through roadshows, engagement at labour nakas and distribution of pamphlets, both in newspapers and through door-to-door visits. Canopies in public spaces are also a general hub for semi-skilled workers looking for work, hence attract attention for the Nirman training. During these activities, Nirman staff members talk about the benefits of the training and motivate unskilled and semi-skilled construction workers to join the course.

The training modules of Nirman consist of theoretical inputs and practical training at construction sites, life skills, safety, basic literacy and numeracy, financial literacy, linkages with government services and schemes and basics of entrepreneurship. The modules are updated according to market needs.

From a need-based and faculty-dependent stage, the curriculum and course structure of Nirman has evolved into a robust, customised and structured programme. Workbooks and reference material based on ITI and similar syllabus help reinforce theoretical and practical learning. Experts teach theoretical concepts in classrooms and, practical learning takes place at live work sites with experienced skilled workers. Courses are conducted in the evening between 5pm to 7pm to ensure that employed workers do not lose income.

Life skills are given priority in the training. Aspects, which ensure continuity in learning, efficiency, growth, and enhanced self-esteem are cultivated during the course. These include understanding identity, motivation, positive thinking and attitude, values and relationships, time-management, communication, decision making and team work. We have observed that life skills have resulted in better jobs and increased incomes for a majority of trainees.

Training in personal and workplace safety is integral to Nirman along with financial literacy to cultivate savings and planning for social security and unforeseen expenses. Procurement of ID documents, opening of bank accounts, linkages with government schemes and benefits, and purchase/renewal of insurance policies is facilitated.

We designed certification of Nirman participants, not only for the skills that they had acquired, but also for the changes that occurred regarding their personal, economic and social development during the training. With Professor Abhishek (he does not use his surname) of IIM, Ahmedabad, we developed an index which measures and compares the changes in ten training components during the period between joining the training programme and its completion. These are:

a. Technical Skills Enhancement
b. Literacy
c. Financial Literacy
d. Access to Government Services
e. Social Security and Insurance
f. Identity/Personal Development
g. Economic Status/Livelihood Enhancement
h. Business/Entrepreneurship Development
i. Social Empowerment
j. Social Mobility

The changes observed during the training are mentioned in the Nirman certificate and serve two purposes. The trainees are empowered because they see the measurable positive change that has taken place due to their efforts and, potential employers feel that they are getting well-rounded personnel.

Women @ Work

Sheetal had always been a homemaker. Her two kids kept her busy. One day, she heard from the women in her colony about Saath's Home Manager training programme. It has been three years since she took up her first assignment and there is no turning back now. "My father-in-law is happy, husband is happy, everyone is happy!" Sheetal says.

She has plans to save enough to be able to buy her own home. She contributes to the household expenses, as well as puts some away towards her dream house.

Deepika, also a Home Manager, was a contrast to Sheetal. Unlike Sheetal's quiet brief sentences, Deepika was bubbly and talkative. She gave up the job when she felt she couldn't cope with the job requirements. Then Devuben, the Saath Coordinator for the project, gave her a pep talk and explained how to deal with the people she was working for. She went back and is still working for them three years later. She speaks very highly of her employers. They are considerate and look out for her. Since she is illiterate, the lady of the house has offered to teach her to read and write. In Deepika's words, "The madam has said she will train me to the extent that I will be capable of doing Devuben's job."

She says she had been a very scared person. Too scared to say anything to anyone, but now is able to communicate with everyone. She is now running the household finances. She has managed to repay a loan and is also saving up for a home. Her in-laws are very proud of her. She attributes her new, confident, happy self to Saath's intervention.

To encourage livelihoods for women, we encourage young women to take up traditional and non-traditional jobs. In two programmes, Saloni and Women@work, which are similar to Nirman, young women are trained to become electricians, drivers, mobile phone repairers, and beauticians. We partner with Godrej Industries in its major CSR programme called Saloni, which aims at enhancing livelihoods by providing training in beauty and hair care. Women trained in the Saloni programme either work in established beauty parlours or become home-based entrepreneurs. Saath is a nodal agency for Saloni in Gujarat and works with 17 NGO partners. We have trained 8,600 young people in the Nirman, Saloni and Women@Work programmes

Livelihood resource centre

A big contradiction in the Indian labour market, especially urban, is that employers complain that they are unable to find suitable workers and, job seekers complain that they are unable to find suitable jobs.

This problem is exacerbated because employers do not mention the specific skills required for a position and job seekers are unaware of the skills required to match the specific job. The aspirations of young job seekers are often not rooted in the reality of the specific market requirements; employers on the other hand, are not willing to go the extra-mile to locate and retain specifically skilled workers, especially at the entry level.

In 2015, we realized the need for a platform where employers could source candidates with specific skills; and simultaneously, job seekers could readily find information on specific skill requirements for a particular job. This would enable them to update their skills, understand and temper their aspirations and, get suitable jobs. We created the platform – the Livelihood Resource Centre (LRC).

The LRC is, a one-stop solution for employers and job seekers. On the one hand, LRC reaches out to employers and registers their entry level HR needs for specific positions with detailed requirements such as, skills, timings, salaries, location of work and job profiles. On the other, LRC has a database of job seekers with their skill profile.

To address requirements by employers for specific roles, a scan of the database reveals all suitable candidates with broadly appropriate skills, who are then shortlisted. Counselors at the LRC interact with the candidate and his/her family to identify key concerns, which include job-profile, location of work, timings and salary.

When the candidate is willing to take up a job, her/his skills are scanned and updated to match the job-profile. This is done through short one-week courses, mainly soft skills. In case the candidate does not have the appropriate job-role skills, s/he has the option to learn these skills from a regular employability-training programme.

The potential candidate then goes through a work-readiness routine, which includes a written assessment as well as preparation for a formal interview.

Once a candidate is selected and placed, LRC provides handholding support, both to the employer, as well as the placed candidate, to ensure effective job-roles and long-term retention.

Apart from its trainers and counselors, LRC has created a peer group to support job seekers. This group helps a young person during the early uncertain, nervous stage of a new job.

Almost 410 jobs have been enabled through LRC

Micro-entrepreneurship

Indian cities and towns see a plethora of individual and family run micro-enterprises in the manufacturing, food and service sectors. These are of food and assorted vendors, small shops, two-wheeler repair, body-care and wellness, pottery, furniture repair and many more. We have listed more than 120 types of micro-enterprises.

The Informal Sector and Micro-entrepreneurship

According to "Women and Men in the Informal Economy: A Statistical Picture", by the International Labour Organisation, published in 2018, **nearly 81 percent of India's employed population is in informal economy.**

The characteristic features of this sector are ease of entry, small scale of operation, local ownership, uncertain legal status, labour intensive, operation using older technology based methods, flexible pricing, less sophisticated packing, absence of a brand name, unavailability of good storage facilities and an effective distribution network, inadequate access to government schemes, finance and government aid, lower entry barriers for employees, a higher proportion of migrants with a lower rate of compensation. Another major challenge the informal sector faces is getting start-up financing. They sometimes lack the skills necessary to manage the financial aspect of their businesses. As a result, many informal sector enterprises cannot grow and develop their business beyond a micro enterprise, while many even fail to earn a livelihood.

In the current scenario of high economic growth with non-commensurate job creation, the informal sector enterprises are one of the main drivers of livelihoods in our cities and towns.

Successful entrepreneurship will not only improve the quality of work and life for low-income entrepreneurial workers, but will lead to job creation, as these enterprises develop and expand.

The people who start and manage these enterprises are hardworking and willing to take risks. Not only are they producers, but also sellers of their products. Many are not professionally qualified and do not have the network to access start-up and working capital. Their constant concern is day-to-day survival. Their level of income is solely dependent on the level of success of their enterprise.

We found it surprising that very few policy inputs or support systems have been made to champion the promotion of the informal sector enterprises. Unlike the formal sector, these enterprises are left to their own devices when it comes to starting, sustaining and scaling up. We do not find the equivalents of IIMs, IITs, incubation and mentor eco-systems with seed and growth funding support for encouraging informal sector entrepreneurs.

India is investing heavily to create skills for the formal sector, which is only 20% of our economy. We have created the National Skills Development Corporation with allied sector skills councils. National and state governments provide generous funding for skilling in the formal sector. Our Start-Up India initiative is directed at the formal sector. The informal sector, we feel is grossly neglected in our national narrative on livelihoods. To change this, we decided to create and demonstrate a model for promoting, sustaining and upgrading the informal sector enterprises – The Business Gym.

The Business Gym model

In the Business Gym model, we work with individuals who operate nano, micro and mini-enterprises. These enterprises are small shops, food, tea and vegetable vendors, cobblers, beauty parlors, street-side cobblers and tailors, etc. We work with new and existing micro-entrepreneurs.

The aims of the Business Gym are to motivate, equip and moderate those wishing to become "nano" entrepreneurs, catalyze "nano" entrepreneurs already in operation, to be part of the mainstream, forecast and model new businesses, help and ensure micro-enterprises achieve higher productivity and profitability.

To design the Business Gym programme, we interacted closely with existing micro-entrepreneurs to understand their challenges and their availability for mentoring and training. We found that the challenges were quite varied for each entrepreneur. They could not spare the time to attend regular classes at the cost of their businesses. It became clear that we could not run a regular training programme. We would have to adopt an individual coaching approach, where a specific need could be addressed. Joint sessions could be held with a set of entrepreneurs with similar needs. What came out strongly was that all the entrepreneurs we met felt that mentoring and coaching was necessary.

To gauge the commitment of participants in the first batch, we decided that we would select entrepreneurs who regularly kept in touch with the Saath mentors over a 3-month period. Out of the 179 micro-entrepreneurs to whom we introduced the Business Gym concept, 94 showed interest. We assessed these participants and short-listed 67 who could participate.

Between March and July 2016, we conducted a detailed baseline survey to collect details on demographics, type of business, turnover, monthly profits, family support, record keeping and accounts, credit facilities, existing marketing strategies, expansion plans, knowledge, access and use of technology, competition, personal and enterprise documentation, communication and negotiating skills, ability to handle challenging situations and readiness to accept mentorship. 49 members were then finalized as Entrepreneur-Members (E-Members) of the program.

We designed a coaching program with seven modules, which were Communication Skills, Teamwork, Business Simulation, Growth Scenarios, Entrepreneurial Motivation/Filtration, Generating Business Plans and Financial Inclusion.

How the Business Gym model works

Business Gym mentors would meet the E-Members regularly and after doing a SWOT of the enterprise, help them understand important business aspects. This consisted of finding and keeping customers, hiring new employees at reasonable wages, finding partners, financing, developing a business strategy and updating it, manufacturing efficiency, setting up accounts and keeping records, sourcing material and input costs at reasonable rates, marketing, accounting, finance, inventory, operations and networking. Coaching and mentoring was based on the requirements of the E-Member and the business. There wasn't a one-size-fits-all type of content and training methodology.

At each interaction, the impact of the guidance given by the mentor would be assessed and built upon. The idea was to let the E-Member implement a new approach, reflect on the effectiveness and adopt, if useful.

All E-Members were aided in understanding the importance of financial and identification documents. This included health insurance, accident insurance, life insurance, IT returns, PAN Card, Aadhar Card, voter ID, bank accounts, recurring and fixed deposits.

Business Gym mentors would provide continuous handholding support to the participants. On a monthly basis, each E-Member was mentored and received inputs as per individual requirement. This included ideal times for maximizing sales, finding and retaining staff, displaying rates to attract customers, identifying new business locations, adding new products, labeling products, sourcing documents, financial inclusion by opening bank accounts, wholesale marketing linkages, technical training and linkages with appropriate regulatory agencies.

We developed a scoring mechanism to evaluate the assistance needed by the E-Member and the sustainability of the enterprise. Ten parameters were identified to be scored on a scale of 1 – 10, with 1 representing a poor score and 10 being the highest

	Parameter	Explanation
1	Level of Participation	This parameter assesses the frequency in which the E-Member kept in touch with their assigned mentor. Frequent interactions got higher scores
2	Regulation	This parameter is based on the regulatory permissions which are required by the enterprise to run its business. Higher scores are accorded to those enterprises which require fewer official permissions and those who obtained the required permissions

Contd...

	Parameter	Explanation
3	Type of infrastructural set-up	This indicator grades the existing operational premises of the entrepreneur. This has been broken down into owned, rented or mobile premises. Higher scores are given to entrepreneurs who own their premises
4	Required capital	This parameter gauges the working capital requirement of the entrepreneur, which ranges from Rs. 0 to Rs. 5 Lakh. Lower capital requirements are given a higher score
5	Confidence and soft skills	The entrepreneur's self-confidence and ability to communicate and negotiate with customers, who in many cases belong to a higher social category, are scored under this parameter.
6	Impact of the training	The level of impact of the entrepreneur's ability and skill from the mentoring and training is rated under this parameter. It also assesses the amount of further training the entrepreneur needs. Well trained and skilled entrepreneurs get a higher score
7	Skill levels	This parameter is based on the skills of the entrepreneur in running his/her enterprise. These skills include marketing, accounting, ability to access credit, inventory management and networking in business. Entrepreneurs with good knowledge on the topic receive a higher score
8	Knowledge of branding	This parameter rates the entrepreneur's ability to create a brand for her/his product or services.
9	Risks associated with Product/Service	The nature of the product or service of the entrepreneur and risks associated with the same, are scored under this parameter. This looks at aspects of risks like internal consumption, human touch, and others. High risk enterprises get lower scores
10	Marketing Skills	This parameter is based on the entrepreneur's ability to explain the uniqueness, qualities or benefits of the product or service.

Learnings from the Business Gym

The Business Gym is probably the first of its kind of intervention to encourage and support micro-entrepreneurs. We want to share our learnings to bring about a wider understanding of the challenge this sector faces.

a. Existing and new micro-entrepreneurs require different inputs
b. New micro-entrepreneurs are willing to attend formal trainings and camps. Existing entrepreneurs prefer counseling at the site of business
c. Record keeping of sales, inventory, expenses and income were unsatisfactory. These aspects need to be strengthened overall
d. Cash flow management and not being able to calculate profit correctly is a prevailing weakness
e. A sense of fatalism prevails regarding regulatory approvals. Micro-entrepreneurs assume that they will not get approvals and preferred bribing, as a regular solution. Similarly, regulatory authorities do not grant approvals as it is not seen as a priority
f. Most of participants were not financially literate. This is a major reason for not being able to access credit for expansion of enterprises
g. New marketing strategies were not readily accepted as they are seen as an expense, which may not result in increased business. This was especially true for e-marketing
h. Awareness of new technologies was limited. At the same time, the willingness to learn about new technologies was absent unless proved useful by others
i. On the job and direct coaching is required to demonstrate the effectiveness of a new strategy
j. Short, medium and long-term budgeting is a major challenge. This results in crisis situations because resources are utilized for social events, which are not foreseen or budgeted for.
k. Sourcing of inventory from wholesalers is a challenge. Entrepreneurs are either not aware of wholesalers or are unable to temporarily shut their enterprises to go to wholesalers
l. Participants were not willing to employ more people out of a fear of theft and under-reporting
m. Levels of education were mainly up to primary and secondary levels, which was an obstacle in understanding course inputs

n. Premises and investment are major determinants. Those who have a family backing or family tradition are at an advantage
o. New micro-entrepreneurs require market linkages and financial assistance. Formal training had a high impact on knowledge levels
p. An absolute absence of formal financing is the biggest hurdle for new micro-entrepreneurs

Social enterprises

We have worked on livelihood enhancement for both, formal and informal sectors. In both sectors, we have training, placement, and when required, working capital and market engagement support.

Our first attempt in creating a livelihood enterprise was not innovative. We designed a sewing and tailoring programme based on the skills we could impart within a limited budget. We started these training classes for women in the slums of Guptanagar in Ahmedabad. We had hoped to enable those trained to be able to find employment in the formal tailoring industry. This did not really work, because the training provided was basic, and the sewing machines used were different from those used in the industry. We could not afford to invest in commercial sewing machines.

The training, however, was useful. The women were able to meet the sewing needs for their families. A few also started sewing for people in their neighbourhoods and began to bring in small sums of money. We found that those who were better skilled and capable were able to take up job-work from larger businesses. After identifying a few contractors, the women started doing job-work from their homes. Through the savings cooperative, they were able to avail of loans to buy sewing machines. About 200 women were trained as tailors in Guptanagar.

This experience taught us that cottage industry type of training for just production, was not adequate in urban areas. The women we trained found it difficult to engage with the market, as they did not have the skills to identify contractors and negotiate fair terms. Their incomes did not increase significantly. Many a time, they could not adhere to quality standards, resulting in rejections and loss of income. **We had leant a** useful lesson. Training and support for livelihoods had to be more robust; it had to be linked with market practices and requirements.

Urmila Home Managers

Our next foray into livelihoods enhancement was a more successful innovation. It started in 2003, when a Saath well-wisher, Ram Kumar, wanted a reliable maid when his wife was expecting a baby. We referred women from the slums we worked in, but he was not satisfied with the quality of services. As Ram is wont to do, he took the bull by the horn and helped design and pilot a professional domestic maid service, which we called the Urmila Home Managers. (Urmila was the mythical wife of Laxman who kept the home fires burning while Ram, Laxman, and Sita were preoccupied being in exile and fighting a war).

Urmila Home Managers broadly had the following objectives.

a. To provide a much-needed domestic maid service with good quality work and reliability.
b. To charge a premium for these services to significantly increase the incomes of women who joined Urmila.
c. To enhance the self-esteem of domestic maids by furnishing them with appropriate certification and skills
d. To create social security mechanisms for domestic maids

To fulfill these objectives, we decided that Urmila should run on the lines of a social enterprise and not a typical NGO income generation programme. We then conducted a market survey to determine what potential clients were expecting from a professional domestic maid service. The survey highlighted security, reliability, and quality as the main requirements for which potential clients were willing to pay a premium.

We then designed a fifteen-day training programme, which would meet these requirements. In principle, training would be imparted by professionals in the particular sector, after which they would receive certificates. Some of the professional sectors that were engaged in the training process were: the Ahmedabad Fire Brigade provided training for fire and safety; St Johns Ambulance, for emergencies and general safety; Service staff from retail stores for operating household appliances. Experts in specific fields provided training – for geriatric, child and patient care, life-skills, communication, and etiquette. Familiarity with cooking methods in with less oil and spices was also a part of the training. Before being deployed, on-the-job training was conducted under the aegis of homemakers who were both, sticklers for quality, and champions of women's empowerment.

To meet security and health concerns, we approached the police to verify the antecedents of the maids and certify that there were no records of criminal offences. This required documents for personal identification and proof of residence, which we helped procure. They also underwent a medical check-up in a reputable hospital, which included relevant tests for communicable diseases, after which they were certified free of any communicable diseases.

The maids were equipped with uniforms, identity cards, kits with an apron, lunch box and water bottle, wrist-watch, chappals, nail cutter, saree pin and a napkin.

The women trained through Urmila, were Home Managers and not domestic maids. Being trained professionally, they were now professional Home Managers, capable and self-confident, in no way the regular maid any more. It also enhanced their self–esteem.

The regular practice of domestic maids was working in as many households as possible to earn a decent income. However, in trying to keep up with her schedule, she needed to complete her tasks in haste, and often ended up compromising on the quality of work, leading to dissatisfaction and recriminations on the part of the employer. Often this would call for a deduction in salary, which did not enhance her self-esteem. Urmila sought to change this practice. Ideally, it meant that a maid would have to work in one house for eight hours or in two houses at the most and be able to earn a higher income than she would have, working in multiple houses.

When the first batch was under training, we spread the word through our network of well wishers. We also informed those households that

that been surveyed about the soon to be available Home Managers. The response was quite good.

We created a system through which interested clients put in their request for a home manager along with the specific type of services required, which could be housekeeping, laundry, child care, cooking or a combination of these. The Urmila staff members would meet the clients in their homes and check out their antecedents, requirements, the size of the household and what would be a reasonable time to complete the required tasks; thus clarifying the job for either an eight-hour, or four-hour shift. If acceptable to both parties, a contract would be signed and a suitable Home Manager employed.

Initially, an Urmila staff member would accompany the Home Manager for a few days to help her settle in, as required by the client. If the first Home Manager was unhappy with the work or with the client, or vice versa, Urmila would send in a second Home Manager. This would work in most cases. If not, Urmila would cancel the contract. We did not offer the Home Manager service to single male houses or where the clients were rude and used offensive language. Occasionally, we received weird requests, such as a fair looking or young Home Manager, which we promptly refused.

Once the Home Manager and client were comfortable with each other and the home care tasks, an Urmila staff member would visit the client once a month to verify the attendance and finalise the number of days, and the fees due. There are two kinds of fees that the client pays; one is the agreed monthly amount to the Home Manager and a 15 percent management fee to Urmila. In case a replacement for the regular Home Manager is required, appropriate fees would be decided and charged. Presently a fee of Rs. 10,000 is charged for eight hours and Rs. 5,800 for five hours.

The perks for the Home Manager include a weekly holiday. The Home Managers are covered under a group accident policy. Urmila helps the Home Manager open a bank account and her fees are paid by cheque. If the location of the client is in an area without public transport connectivity, conveyance expenses are paid additionally.

Urmila was modeled as a social enterprise, with the goal of being financially self-sustainable, which it has achieved.

Over the years, Urmila has deployed about 1,000 Home Managers. **These women have acquired skills and certification and are self-confident. Their aspirations have increased, and they want a better life for themselves and their families.** They are educating their children and their social status has gone up. They are no longer domestic servants. An unintended, outcome is that the Home Managers are gentrified.

RWeaves – bringing back arts and crafts

The earthquake of 2001 took us to Surendranagar district where we were working on the Sustained Nutrition, Education, Health and Livelihoods Project (SNEHAL) supported by Care India. This aimed at enabling livelihood security for the local workers and micro enterprises.

It was during this time that we came across a couple of Tangaliya and Patola weavers. These are a very specialized art of weaving. Patola is a very fine silk weave with traditional designs. Tangaliya too has its own patterns and designs. It was traditionally, however, woven only in wool. The tragedy was that both arts were on their way out, especially the Tangaliya weaving.

We decided to help revive the Tangaliya weaving. There were just two weavers working at it. Initially, we bought their stock of material and marketed it within our organization and among friends. But that was not all the help they required. With the professional help from the National Institute of Fashion Design (NIFT) we held a training programme to introduce cotton into their weaving, as well as use of different colours and patterns. It was attended by most of the original weavers. We got a Geographical Indicator (GI) tag for Tangaliya.

After the training, once again, most of the weavers went back to the trade that they had taken up when they were not able to market their produce. It was an uphill task encouraging them to go back to weaving. To enable them start once again we assisted them by introducing them to

Rang De, an organization that helped artisans through loans. Gradually, fourty weavers returned to the art.

We assisted in expanding their line of products, as well as that of the Patola weavers. To market both products, we developed the brand name "Rweaves." **Today, there are approximately fifteen more families and sixty weavers weaving Tangaliya once more.** The Tangaliya and Patola products are sold through Rweaves, through exhibitions nationally, and online internationally. Locally, we stock and sell the products from our premises on a regular basis. As word has spread, many people visit the artisans to see their work first-hand. They now receive specific, personalized orders as well. We are hopeful now that the art is a well-paying livelihood, many more will get back into the fold.

Chandubhai Rathod from Dedadra village, the epicenter of Tangaliya weaving, in Surendranagar District, is a contented and happy Tangaliya weaver.

He recalls, how over the years, the use of woolen material decreased as the colourful, mill-spun, cotton material took over. The Rabari, Maldhari, and Bharward tribes (shepherds and nomads) remained their regular customers. Due to a drastic drop in the demand for the material, it became difficult for them to earn a decent living. Most of the weavers turned to other trades. Tangaliya weaving became a dying art. For some of them though, it was an art that they refused to let go of and, did their best to eke out a living. Eventually, three weavers stuck to their trade.

Chandubhai smiles broadly, when he talks about when things took a turn for the better. People from Saath, NIFT (National Institute of Fashion Technology) and CARE Interntional, an organisation from Delhi had come looking for Tangaliya weavers. They prepared a project on Tangaliya weaving. As part of the project, weavers underwent training; they were taught to weave in colourful cotton and silk, instead of wool. They were exposed to marketing and sales. An exhibition was organised, which went off very well.

Once the project with the three organizations was completed, the weavers were left to continue the work on their own. Even though they had been exposed to ways of sales and marketing, they found that they were unable to deal with that aspect of the work.

Since they had gotten to know Bellaben during the training period, they contacted her once again, with the problem of selling their products. Initially, Saath started buying the products, as well as selling them. They also took several visitors to experience, first hand, the art of Tangaliya weaving and meet the weaving community.

Bellaben was very encouraging and informed them that Saath would continue to help with the marketing of their products. She said they would do everything to market Tangaliya products at national, as well as international level.

"True to her word, our product is in the national and international markets through web portals," Chandubhai beams.

He said they have come a long way with regard to their income. When the art started dying out, they were just about able to meet their household expenses. For any expenses towards celebrating functions or festivals, they would go to moneylenders for loans. They no longer need to take loans; they are financially stable now.

Chandubhai says, "Since Saath's intervention, we have no difficulty with the marketing and sale of our products. We are also able to avail of loans, whenever needed, through Saath from Rang De. This is especially useful, as we do not have to run around for the process of obtaining a loan, like we have to for loans from banks."

On enquiring as to the number of people from his family in the trade, he replied, "My brother, three cousins of mine, and recently, my son has also taken up this work." As to why he had only been counting the men folk in the weaving process, he clarified that women do the process preceding the actual weaving. They prepare the reels and bobbins that go on to the looms to start the weaving process. The women are an integral part of the process and play a very important role. They also assist in the warping process, which cannot be done by a single person. There are also a few women who weave.

Thirty weavers were trained, but only fifteen are working. He feels with more hands they will be able to service large orders more efficiently, as well as increase regular production. According to him, his efforts at encouraging them to come back to their original livelihoods yielded fifteen looms. He has argued with them about the wisdom of having to work according to other people's schedules at jobs they had to learn, against the freedom to work at their own time and pace at something they had been doing for years. He has also argued about the amount they could earn would be up to them, in the number of hours they put in. He further said that there were a few more who could be brought into the fold, as they did seem to be convinced, but did not have any of the equipment. They would probably need loans for the purpose.

CHAPTER 6

FINANCIAL INCLUSION

Sudha Rawal has been living in the Vishali Towers, Vasana, Ahmedabad, for the past seventeen years. She has a business selling tea leaves. In the beginning, she ran her business from home on a very small scale, personally delivering orders to people's homes.

In 2004, she met Rekhaben from the Saath Credit and Savings Cooperative, who told her about the savings and loan schemes. She opened an account in the Saath savings scheme when the compulsory saving amount was Rs. 100. Her first loan from the Saath Cooperative was for Rs. 5,000. Today, she has taken a loan of Rs. 1,00,000. Most of the loans were for her business. The last one is for the renovation of her house.

With the help of ready money, Sudha could grow her business. Instead of having to personally deliver the tealeaves to her customers, they now came to her for their requirements. Her compulsory savings enabled her to send her children for tuitions, when they were not doing too well in school. She was also able to buy household furniture.

Then there is Pinky Rathod – a married woman with two children. She too has been living in Vasna for the past sixteen years in a small one-room kitchen flat. When she came across Saath's compulsory savings scheme and loans programme thirteen years ago she immediately grabbed the opportunity. Pinky had been living a hand to mouth existence, and so saving even Rs 100 a month was very difficult. However, she made her payments regularly and was able to avail of her first loan of Rs. 5,000 and then subsequently, over the years, larger amounts. She now is repaying her Rs. 1,00,000 loan.

She started out selling small quantities of readymade garments under a small canopy. Her husband produced agarbattis manually. Today, she has a stall of her own and her husband is the owner of an agarbatti-making machine. They have been able to acquire all household gadgets and equipment like, an air conditioner, washing machine, an Aquaguard, fridge, etc. "The loans from Saath came in very handy when my business was not doing too well," she said.

This is a classic case study of how small loans can considerably enable a person or a small business to spread their wings.

Financial inclusion is imperative for the economic growth of individuals, families and a country. In India, financial inclusion has been hampered by lack of awareness, penetration of formal banking institutions, onerous procedures and documentation, and attitudes of banking staff towards vulnerable informal sector clients.

According to the World Bank Report on Financial Inclusion 2018, 79.8% of adults in India have a financial institutional account. The report also states that 38.7% of account holders did not make a deposit or withdrawal from a Financial Institution account. The Jan Dhan Yojana, a government initiative towards financial inclusion witnessed a large number of bank accounts being opened, but a significant number of these accounts remain inoperative.

People engaged in the informal sector have been historically excluded from formal financial institutions. They may be able to open bank accounts as individuals but getting credit and working capital for personal and business purposes is a challenge. The gap was filled by private moneylenders who charged usurious interest rates. After Prof. Mohamed Yunus demonstrated in 1976, through his efforts in the Grameen Bank in Bangladesh that the poor are bankable, a number of micro-finances started in India to provide financial services to the informal sector. The Micro Finance sector has grown by leaps and bounds in India and has played a significant role in ensuring financial inclusion in both, rural and urban India.

Ekta Bachat Mandal

Our endeavors at Saath towards financial inclusion started in 1990, when we started a revolving loan program to construct toilets in

Sankalchand ni Chali (SMC) in Ahmedabad. Residents who availed the loan had to open bank accounts to deposit cheques they would get as subsidy from the government. We found that it was almost impossible to open accounts because the staff was not keen on getting customers who would keep minimal amounts as savings and rarely able to invest in fixed deposits. Even in those days, objections were raised regarding the KYC documents. Ultimately, we had to become joint account holders to open bank accounts.

Many youth in SMC were micro-entrepreneurs, who used to buy clothing items in bulk in Ahmedabad and travel all over the country to retail these items. We found that for working capital, they took loans from their suppliers at interest rates, which were as high as 60 percent. It was impossible to get loans from banks. We then supported the youth in starting a self-help group whose members would save amounts ranging between Rs 50 – Rs 100 regularly and become liable to take loans for working capital for their micro-enterprises. Loans were also given for medical and other emergencies. By 2007, the Ekta Savings group had 2,450 members

Sakhi Bachat Mandal

When we started working in Guptanagar in 1992, we found that residents faced a similar precarious financial situation. Though people worked and managed their households on their incomes, very few had ever saved or even thought of saving. Whenever the need arose for an unexpected expense, or even a planned one, they pawned jewelry or took money from moneylenders at exorbitant rates of interest. Repayments were a nightmare for them, as they could barely repay the interest installments, which led to the debt increasing.

In 1996, when the Slum Networking Project (SNP) was launched in Guptanagar, residents had to contribute Rs. 2,000 per household as their contribution. The women in the community suggested that they could save small amounts on a monthly basis towards their contribution. We held a series of meetings with the seven community groups that had been formed to facilitate SNP to explore the idea of forming a women's savings group. During these meetings the women were asked about their financial status in terms of what their incomes and budgets were. It was then suggested that they could start saving a rupee a day, thirty rupees a month. Back then, approximately twenty-two years ago, saving Rs. 30 a month was a big step. They decided to cut it down to Rs. 25 a month; and our women's micro finance sector was born.

Initially, we started with just thirty-five women, which grew to around 100 members, in all. Responsibilities for each area were then handed over to the Saath women overseeing various activities in the area. After which, women from the area, who were literate, were given the responsibility of collection and keeping an account of the savings. They made all required entries in a register, as well as in the client's passbook. The process being handled by the local women gave them a feeling of security where their money was concerned. On the 10th of every month the money was then taken to the Saath office, where once again, it was counted and entered into a register. This was then deposited under Sakhi Mahila Bachat Mandal into the Sakhi Mahila Mandal account, which was operative at that time in the SEWA Bank. This process ensured there was accountability at each stage.

A couple of years down the line, there was a request for increasing the savings amount. However, this was not a unanimous request. It was then that separate saving categories were put in place – Rs. 25, Rs. 50, and Rs. 100.

In 1999, our membership went up to around 300, and our corpus had reached a lakh rupees. We then decided to diversify into disbursing loans. The monthly interest rate was 1.5 percent. The ground rule regarding the loan amount to be sanctioned individually was one could only avail of a loan amount not more than three times the amount saved. There had to be two guarantors to safeguard repayments, and further, they could not avail of a loan or draw money from their savings until the existing loan was repaid. Initial loan requests were for Rs. 1,000 which gradually increased over the years.

We had started working with residents of Saraspur, Behrampura (Santoshnagar, Mohan Darji ni Chali and Jethalal ni chali), and

Sankalitnagar who were severely affected by the 2002 Gujarat riots. Care India, as part of its rehabilitation efforts had given subsidies to the affected families for restoring livelihoods and housing. When a degree of normalcy was restored, we explored the possibility of starting savings and credit groups in these areas. We started with a group of 35 persons in Berhampura. In the Ekta Bachat Mandal, by 2005 we had a membership of 1,000 from all the three slums. However, 20 percent of them were very irregular. Similarly, in Saraspur, a group with twenty members was formed and increased to 500 members by 2005.

In Sankalitnagar, Juhapura, residents were initially suspicious about Saath's intentions. Yakubbhai Pathan, a resident of Sankalitnagar, who had known Rajendra Joshi, vouched for Saath's credibility and, eventually a savings group of 40 women was formed. Women would deposit amounts daily and initially would withdraw the entire saved amount. As confidence grew, they retained their savings and started taking small loans.

In 2002, both the savings groups were registered as savings and credit cooperative societies. Sakhi Bachat Mandal had women members only and covered Vasna, Paldi, Juhapura, and Saraspur areas, while Ekta Bachat Mandal covered Behrampura – Santoshnagar, Jethalal ni chali and Mohan Darji ni chali and was a mixed group of men and women.

In 2007, Care India stepped in and suggested that the savings and credit programme be run on a professional micro finance basis, as the combined membership of the two bachat mandals had grown to about 3,500 members.

Saath Savings and Credit Cooperative Society

We then brought in consultants on micro finance, Access Development Services from Delhi, who periodically visited us, and were of the opinion that our existing terms for giving loans were rather daunting. We then held discussions and designed and formalized a new system in which members would have to make compulsory savings of Rs. 100/month for six months, after which, they were liable for loans. We then introduced the Joint Liability Group (JLG) practice. In the JLG process, members formed a group ranging between four to seven members. They stand as guarantors for each other in the absence of any collateral.

We were the first micro finance institution in Gujarat to start the Joint Liability Group. It was already in practice in Madhya Pradesh and South India. We had a very tough time explaining the concept in a positive way. People felt it was a very difficult proposition. They found it difficult to trust their close family and friends in money matters, forming a group of trustworthy persons almost seemed like an impossibility.

Though we understood the concept, it was exceedingly difficult to make it a reality. In September 2007, our micro finance group visited Basix, an organization in Indore, which was successfully operating the scheme there. After the visit, we were confident about running the programme. We held meetings in each area and were able to explain and clarify the scheme. We felt we had good prospects, as we were the only other micro finance institution, after SEWA Bank, in Gujarat.

To introduce the Joint Liability Group loan scheme, we had to upgrade our software and started keeping our Management Information System records. Training followed, as were only used to working in Tally, the simple accounting software. We were able to do all of this because of the funding support from CARE India. Our consultants from Delhi who initially visited us once a month, in the following year, 2008, visited bi-monthly and then every quarter in the next year.

Through this transition, we also built the micro finance organization structure – a CEO, Branch Managers, Accountants, and Field Officers. This was possible as we worked as a cohesive team, learning and teaching each other.

Disbursing loans had always been a very tough call. There was no way to streamline checks and balances in the informal sector. You could never be sure that the person applying for a loan did not already have an outstanding loan elsewhere. Taking loans from moneylenders was the norm. We developed an appraisal form, which included details of family, income, home appliances, if in business, available stock, etc. This helped tremendously and the process was streamlined, to some extent, and the loan sector started growing. Recovery rates improved too.

We piloted the scheme a year later, in 2008, in Vasna and Juhapura areas. Subsequently, it was introduced in all the other areas.

In 2007, we covered only seven to eight areas. We had to seek permission from the government to operate a cooperative in any new area. We expanded to twenty areas. We also started the process of merging Ekta and Sakhi Bachat Cooperatives for better management and streamlining processes. It was for the first time that two profit-making cooperatives were merging. It took the Registrar of Cooperatives, Gujarat, three years to approve the merger. In 2010, we received permission and the merged entity, called Saath Credit and Savings Cooperative Ltd. was launched.

After the merger, we encountered largely technical problems, not so much operational. Audit of the cooperative was one of the major hurdles we had to face. However, for the first three years, the government audit did not take place. In the third year the statutory government auditors arrived and had a difficult task of sorting and understanding the merger. One of the issues was, understanding the inclusion of the account of the Community Based Organisation (CBO) that was also merged into the Saath cooperative. As the merger took place on 19th March 2010, the entities were audited individually up to that date, the two registered cooperatives and the CBO. These audits had to be merged into a single audit report after the merger, i.e. 19th March 2010. This particular process was extremely tough. The statutory auditors who finally came calling in 2013 took almost a year to complete the audit.

Looking back, we felt that we were better off when we had prepared our 5-year plan in 2007. In 2009 we had four branches in Vasna, Behrampura, Saraspur, Juhapura. All expansion of branches happened after 2010.

In 2012–13, we started Vadaj and Sarkhej branches, 2013–14 Shyamal branch which is a virtual branch for the investors, 2014–15 Bareja, Hathijan, Gomptipur, 2015–16 Naroda, 2016–17 Ode.

During the period where our consultants assisted us, our products included, compulsory savings, Joint Liability Group loans, voluntary savings, and the fixed deposit scheme. As the demand for these products increased, we started facing a cash crunch. A UK based organisation, Shivia, stepped in with a zero-interest loan through Saath. The fund was of immense help. However, after changing over to the Joint Liability Group loan system, where each member of the group availed of a loan, the demand increased. After Shivia, we also took two more loans from IGS (Indian Grameen Services) at an interest of 15 percent per annum for the value of 30 lakhs and 50 lakhs, which helped with liquidity.

We then started a Child Plan savings scheme with a compulsory deposit of Rs. 200/month, which was the only scheme not linked to loans. Although it did not take off as expected, it did moderately well. Further, we introduced the Rs. 500 monthly recurring scheme. This too had a moderate rate of success. Our earlier loans had been on the basis of the amount of compulsory savings the client had. We now give loans against any scheme they invest in, except the Child Plan Scheme. The minimum amount was around Rs. 3,000 and maximum was Rs. 30,000.

Our attitude towards the Joint Liability Group loans started to get a bit lax. As all of the clients were guarantors as well, we didn't envisage much of a problem in repayments. Further, the amount of loans too was larger, in most cases. When clients who had been with us for ten years and more approached us, we tended to clear their loans, no matter what the size. This was our downfall. Disbursing bigger loans to old clients, without proper loan appraisal, made loan recovery difficult. Our loan debts rose. Ten percent defaulted on repayments. We then asked Rajendrabhai to help convince older clients to repay the large loans. This system seemed to work, and our recovery rate went up.

We then launched the Monthly Income Scheme. On a deposit of Rs. 10,000 the client received Rs. 80 every month. The scheme was a moderate success. **But our enthusiasm knew no bounds and, we introduced the "double your money in six years" scheme the next year, along with some good incentives. It was a huge success; in the very first year we received more than 40 lakhs.** This overshadowed the

Monthly Income Scheme. Our clienteles dwindled to five to six persons a year. We nevertheless decided to keep the scheme going.

In 2015, we introduced the "shopkeeper loan" scheme, a single person loan, Rs. 50,000 and above. However, we decided on the criteria that the person should have a record of three prior loans from us. This backfired, and we got just two cases in the year. We did a rethink and changed the criteria and we had a better result in the following year, twenty-five loans were disbursed.

One of our worst experiences during the year was from within our organization. A field officer, who was very well acquainted with the clientele collected the installments but would tell them that he had forgotten his receipt book and would give them a receipt later. Further, he would borrow a few thousand from those he knew had the money. This went on undetected for a period of about six months. All of it came to light when the clients approached us for a loan during the Diwali season. On being told that they had outstanding installments, they denied it saying that they had paid all their installments but had not received a receipt from the field officer. This was an eye opener and we realized we had to be more vigilant. We have an ongoing case against the field officer.

To counter the problem, we started our customer care service and obtained a toll-free number. This was also followed with promotional SMS messages that money was not to be given without a receipt, if done, it would be their responsibility also started transactional SMS message system, so for every transaction the client will receive an SMS from the cooperative. The toll-free number turned out to be very useful. Through the calls received we are able to sort out issues faster.

In 2016, when the ban on 2,000- and 500-rupee notes was being implemented we were at sea; there was no directive as to how it should be handled at our end. This was during the Diwali holidays so most of our work had shut down for the festival. In the Muslim areas however, the offices were functioning, and the currency was being accepted. Ours were mainly cash transactions – installments were paid in cash. We stopped all transactions for a couple of days. But it affected our collection severely. Since everything was so uncertain, and experts were giving conflicting advice, we decided to start accepting the banned currency for a period of three days. We were able to recover 80 percent of the loan installments within that period. When we deposited the money into the mainstream bank, they sealed our account stating large deposits of demonetized currency, as the reason. But we were able to get by through our other accounts until such time we were able to have the issue resolved.

On the bright side, we were able to disburse loans, as all loans were made directly into the client's account by transfer or cheque only. Since all loan clients had to have bank accounts, repayments could be routed through their account directly or by cheque. For those who did not have cheque books, the installments were paid through an account holder who had the facility. Another good outcome was that though some of them had cheque books, they had never used them, as they did not know how to sign. During this period, they learnt to sign and it became the norm to pay by cheque. People became literate regarding bank procedures too.

One issue that surfaced was that most of them held Jan Dhan accounts, which had a limit of Rs. 10,000 on transactions during the month. These accounts were closed, and a general savings account opened. This was a tiresome exercise as it took up a lot of time of the client, field officer, as well as of the branch manager. Once again, there were issues regarding compulsory balance required in the account. Sometimes payments through ECS/NSDH were held up due to lack of the required balance, the client had forgotten to deposit the money in time. But these were issues that were gradually sorted out.

Today, in 2018, out of 26,000 clients we have 3,500 clients using ECS for their transactions, and 9,000 clients availing of loans. On an average, we receive 75 percent of the collection regularly.

Our growth path though uneven has been good given all the circumstances. Between 2007 and 2013 it was a very encouraging 20–25 percent each year.

In 2014, after the issue with the field officer, we decided to have the clients come to the office and make their payments. We felt it would serve multiple purposes – to repay the loan, to visit the office, and meet and interact with the people associated with the project. They would also have the opportunity to find out about all the products offered. And to add to it, we had spent a lot of money in trying to make it function well. We increased the staff, made additional arrangements for seating, bought cash counting machines, and opened sub offices where the nearest branch was too far away.

It all seemed so idealistic in theory, but practically it was close to a disaster. In a span of three months our repayments dropped drastically. In those three months 1,800 to 2,000 clients gradually closed their accounts. Most of those who completed repaying their loans closed their accounts.

We all learned a very important lesson; drastic changes in operations all at once, do not work, they backfire. There should always be options. We reversed the decision after four months. The damage, however, was done. We suffered a loss that year. Although we broke even the next year, it took us another year to get back on our feet.

In 2017 we had a clear profit.

Ten urban and two rural, full-fledged branches are functioning today. We have been able to stay afloat with assistance. To become self-sufficient, we decided to create our own core fund, rather than have to take loans. In an effort to become self-sufficient, we have opened a branch which caters to investors, who generally do not need loans. This forms our core fund.

We had started out with loans between 3,000 and 30,000. Today, our loans start at 30,000 and the upper limit is 2,00,000, 50 percent are individual loans. We are now looking to convert people who are just into saving to become investors – for the future of their families.

There has been a very palpable growth. Those who had started saving with Rs. 25, went on to save Rs. 50, 100, 200 then graduated to 10,000 fixed deposits and later, to double in six years. They have also invested in the child saving scheme. We also have many clients investing in multiple schemes, and some have savings up to Rs. 50,000.

The Sakhi Savings and Credit Cooperative Society

The Sakhi Savings and Credit Cooperative Society had its roots in a child rights program supported by IKEA and Save the Children.

IKEA, whose products include cotton material, had been accused of buying cotton from places where children were used for labour, mainly in India and Pakistan. Their products were being boycotted. In 2009, we started a three-year project from Save the Children Fund supported by IKEA on child rights. This work was to be carried out in the cotton growing areas of Dholka and Viramgam. We had identified that the causes for children working in the field were primarily because of health issues, dropping out of school, inability to sustain education costs, and income issues.

This project comprised of four interconnecting components: child rights, income generation, livelihoods, and women's empowerment. Our interventions in the first year brought out deep-seated issues in each of the components that we dealt with.

Working with the women self-help groups, we started the savings programme, but hit a roadblock when the banks refused to open group accounts. However, the women were very insistent on a savings platform in a safe and secure way. We then had discussions on the issue within Saath and felt that a long-term engagement would be possible through such a savings platform.

The project coordinator decided to take on the responsibility and the savings programme was initiated in January of 2011. The money collected were not very large sums, however, depositing the money in a bank was still a problem. Finally, a short-term solution was found. A Saath Livelihoods Services account was opened into which the money was deposited. By 2012, the self-help groups had saved about Rs. 25 lakhs and loans amounting to Rs. 22 lakhs had been disbursed. The local Saath staff in the project had helped create trust and credibility.

When the Child Rights project ended, we debated on the continuation of the savings groups. One option was to merge these groups with the Saath Savings and Credit Society, which had an urban focus. However, the administrative costs would be very high for the urban staff to cover the rural areas, approximately 50 kilometers away. Getting to know the clients would have to start from scratch, which also brought out the problem of the clients having to build their trust in them. We decided to form another cooperative society.

In June 2012, the Saath Mahila Savings and Credit Cooperative Society (SMSCCS) was registered as a women's cooperative society. The accounts were transferred to the SMSCCS account.

By 2013, the corpus had reached 80 lakhs, run and supervised by a team of five members. By this time the Child Rights project had reached completion and we had to terminate the services of the project staff, except the five who were employed by SMCCS. Some of the terminated staff started spreading rumors that with Saath withdrawing from the project there was no security as far as their funds were concerned. They went on to incite the clients who had taken loans not to repay them. This led to an exodus of clients from the cooperative.

It took us six months to individually meet each of the two thousand five hundred clients and explain their money was as safe as when the Saath project was running. The money was returned to those who were not convinced. Among those who had taken loans, there were defaulters. Our client membership came down to around twelve hundred clients.

2013 was a difficult year for us. Apart from rebuilding trust with members, we had to ensure that salaries were paid, new software installed, and processes were streamlined. We managed to get a loan of 60 lakhs from YES Bank on business correspondent terms through Saath Livelihood Services to tide us through the financial crisis. Slowly, but surely, we got back on our feet. We worked out a plan to have 5 branches within a fifteen-kilometer radius to streamline the operations. We further decided that compulsory savings for loans would no longer be criteria. In 2014, we partnered with Basix, from whom we took a Rs. 20-lakh loan.

However, our first venture into a completely new area did not work out. There were many micro finance organizations in the area, which disbursed loans without the need for a savings account with them.

Cooperatives are generally blacklisted by banks and other external financial organizations. There is a belief that cooperatives are mismanaged, there is no accountability because they are run by illiterate people, who do not understand the formal way of functioning. After the initial loan from Basix, it was very difficult to obtain external funds. Our corpus wasn't large enough. Further, we had just one member from Saath on the board. The rest of the members were local. Funders who did come forward found that the committee members were not able to convince them or give satisfactory information.

By early 2015, our corpus grew to one crore. We were able to avail of external funds to the tune of 25 lakhs from Gruh Finance, which was a confidence booster.

In 2016, we opened a branch in Patdi village in Surendranagar district after ascertaining if it could work. The Viramgam staff members visited Patdi and organized the group and collected the money. This continued for four months, as they were regular in payments. We also started operations in Kadi, where we found that there was a demand.

In 2017, we then approached the registrar for permission to operate in four districts – Surendranagar, Mehsana, Gandhinagar, and Kheda. Once we had obtained the permission for these four districts the ADC Bank and Gruh Finance supported us; Basix gave us a second loan as well. Our overheads have been low.

We have never had to look back since. We started with one branch and, today have six fully functioning branches. The project coordinator was experienced in forming Self Help Groups, their functioning and dynamics. He had studied the urban cooperative and its workings and problems and was able to steer clear of them.

There has been a bias against rural savings; it was considered to be irregular, as the source of income was largely dependent on agriculture. Agriculture is dependent on many factors, which could lead to loss of work and income too. Our experience, however, has been that people in rural areas are much more loyal compared to the urban populations. In the villages, there is a lot of social pressure. If a person has defaulted in a loan, the neighbour will warn you about it. In the urban setting this does not happen, the neighbour might not divulge whether his neighbour owns the house or is rented. In a village, you will hear about the village, as a whole.

All our branches are showing healthy profit margins. Our net profit in 2017 was 26 lakhs.

With our success in micro finance with women, we have started women related social activities involving livelihoods training.

"The Loan that Changed My Life"

"My name is Jameelabanu Abdul Razakbhai. I used to live in Tai Wadi, Viramgam. My family had shifted to Ahmedabad for a while. When I was ten years old, we moved back here and continued our business of making and selling bhegatia and golas in a cart. We also did some tailoring work.

"I got married in 1985 and moved to Bhati Fali. I started working as a tailor in the area, I worked long hours, but people did not pay up immediately. It wasn't lucrative. My husband ran a light provision store (Paan Galla) in the marketplace. But it did not do very well. After the riots all businesses took a hit. We have two sons. When my younger son was ten years old, along with the tailoring work, I started this business of making and selling bhegatia, pani puri, pakodis, chicki, puffs, and bottles of flavoured syrups. This I sold from my house.

"Four years ago, I met Kishorbhai who told me about the savings and credit schemes of the Saath Cooperative. I joined the cooperative and took my first loan of Rs. 10,000, which went a long way in procuring raw material. I didn't have to buy on credit any more. Once I had repaid this amount, I went on to take a Rs. 20,000 loan and then a 40,000 one. Today I am repaying a Rs. 50,000 loan. This has been a big boon to me, as I do not have to go out for loans.

"Before I started taking the loans, I bought all my raw material on credit. When I paid the creditor the entire amount I had on credit, I could only then buy my next stock of raw material. There was always the stress of putting together the required amount to be able to get the next lot of material. Now, it is so much easier, I buy all the ingredients in cash; I am able to repay the amount in easy installments.

"My business has grown. After I took the first loan of Rs. 10,000 we rented a shop in the marketplace. Besides, the Paan Galla my husband has been running, my son mans the other outlet that sells all the products I make at home. In effect, I now have two outlets; women come to my house to buy the products while the men go to my son's outlet. Our eateries open at around 5 in the evening. I deliver the items to my son's outlet at around 4.45 and clear up the house to get it ready for the customers.

"Another advantage of joining the cooperative is that I am also saving money regularly with them. I am now in a position where I can run my home without monetary constraints, have a savings account and, can avail of a loan whenever the need arises. I have been able to repay each loan without defaulting in a single installment.

"I am very happy with the assistance from the Saath Cooperative, which has tremendously helped the growth of my business over the past four years."

CHAPTER 7

How to Channelize and Empower People

The term 'community development' can have different meanings for different people. Uniform development of a whole community, catalysing communities towards development, creating or supporting leaders in a community so that they can lead a development process. Community development can conjure images in which a community spontaneously rises to fight injustice with firebrand leaders.

In reality, 'community development' is a time-consuming process of making a community aware, identifying and nurturing leaders with community values, and empowering them how to work together towards a common goal. Most importantly, community development rarely happens through motivational talks and lectures. It requires a set of long term-activities in which the community participates leading to a common good.

How sports brought about community development

Saath's first engagement with community development started in May 1989, with the Sankalchand Mukhi ni Chali (SMC) community in the Behrampura ward of Ahmedabad. The chawl had originally been built to house workers in the then booming textile industry in Ahmedabad. With the decline of the textile mills, unemployment rose, and Sankalchand Mukhi ni Chali became overcrowded, with decrepit houses, and a stressed drainage system. There were only twenty public toilets for 400 households. Water was available at common water posts. The residents comprised of two communities, one which hailed from the Jalore and Nagor districts of the neighbouring state of Rajasthan, popularly known as the "Saragra Marwadis" and, Dalits from the Saurashtra region of Gujarat. These two communities hardly interacted with each other.

The co-founders of Saath decided to work in SMC, not only because it was a stressed chawl, but also because we had had some contact with the Saragra Marwadi community while working with another NGO in a different chawl. The experience helped gain inroads into the SMC. The first step was being introduced to the existing leaders in the chawl. These were traditional community, as well as economic, and political leaders.

Our goal was to create a group of young people who could lead community development processes in SMC. Our priority was to establish a relationship of trust with the youth. We decided to start up the relationship through sports, an activity that would interest the youth. But what kind of sports? Outside SMC, the Ahmedabad Municipal Corporation had built the Liladhar Bhatt community hall, which had open space, where the youth played a kind of children's cricket. We decided to introduce volleyball, as the space was enough for a standard volleyball court. They were not so keen as they did not want outsiders disturbing their game of cricket. We negotiated with them. Ultimately, they agreed to play volleyball thrice a week, in the evenings.

We bought a ball and a net; they contributed a set of poles to tie the net to. One of the young men who was familiar with volleyball, helped mark out a standard court and agreed to be the coach. Inroads to community development through volleyball had begun. Initially, we could not form two teams, as there weren't enough players. The Saath staff of two joined in and became active players. Gradually, more young men from SMC joined in.

We had achieved our goal – we gained their trust. Gradually, we were able to reach out and get them to talk about their aspirations, living conditions, and the challenges they faced. As with young people anywhere in the world, they wanted to have a better future for themselves and their families. They just did not know how to go about achieving it. We explained Saath's objective was to enhance the quality of life of those willing to make an effort, and it was possible, if we worked together as partners, to bring about change in SMC. The youth committed themselves to the partnership. Arriving at this stage, took almost three months.

The first task was to create a sense of purpose and comradeship between the youth. We held three-day camps, where everyone stayed in tents, and cooked their own food. The camp activities were structured so as to give the youth leadership roles – taking care of and managing various activities, which included handling discussion sessions, games, housekeeping, kitchen chores, and entertainment.

Typically, about thirty youth from both, Saragra and Dalit communities participated in the camp. Discussions revolved around how they could bring about changes on issues relating to education, employment and livelihoods, restrictive social norms, and lack of sanitation. In August 1989, during the first camp held, we, together with the youth, decided to get accurate socio-economic information about SMC to aid in understanding the challenges and help create an action plan.

The socio-economic survey was jointly conducted and analysed. This survey indicated four priorities for SMC – lack of sanitation, high incidence of tuberculosis, a lack of working capital for a majority of youth, whose livelihood consisted of micro-entrepreneurship, and a high dropout rate among school going children.

We decided to focus on the sanitation situation first. This was done by plugging into the 80:20 toilet construction scheme of the Ahmedabad Municipal Corporation (AMC)[1]. Involvement of the youth was pivotal. First and foremost, they interacted with key officials in the Ahmedabad Municipal Corporation to understand the requirements of the scheme. They then identified and convinced the first batch of twenty families to take a loan to build household toilets. They further assisted with the official applications to the AMC in prescribed formats with supporting documents, identified and supervised the contractors

---

[1] Described in Chapter 1, Page 20

to build the toilets as per scheme specifications. They helped open bank accounts for applicants, processed the loan documents and ensured repayment. They finally submitted completion certificates for the 80 percent reimbursement from the AMC. Saath's role was to provide guidance and support, when faced with bottlenecks.

Through the hands-on experience, the young people who had participated in the toilet construction programme were transformed into leaders. They had learnt how to work as a team to complement strengths and weaknesses. They overcame their differences and worked with both communities. Their self-worth and confidence had grown with their success in changing the sanitation conditions in SMC. They gained the trust of the community and became role models, a crucial requirement for equitable community development.

They decided to formalize the group with the objective of bettering the quality of life in SMC. The Ekta Yuvak Mandal was registered.

The youth were now confident and ready to tackle the high incidence of tuberculosis in SMC[2]. The role of the members of the Ekta Yuvak Mandal was to identify probable TB patients, convince doctors at the Ahmedabad TB Hospital to diagnose and treat them, ensure the patients regularly took their medication for the prescribed period. They were also responsible for distributing nutritional support. The incidence of TB in SMC reduced and youth from Ekta Yuvak Mandal could take credit. Once again, their success in dealing with the TB issue reinforced their self-confidence as leaders.

For easing access to working capital for micro-entrepreneurs in the community, Ekta Yuvak Mandal promoted a savings and credit cooperative called Ekta Bachat Mandal[3]. The initial role of the youth was to find out how a cooperative could be formed and registered, understanding the rules and bye-laws, convincing residents to save, working out norms for loan disbursements and ensuring repayment.

---

[2] Described in Chapter 8, Page 96
[3] Described in Chapter 6, Page 73

Through the process of learning all about cooperatives and, then registering one, they were able to successfully run it.

Did community development happen in SMC? We believe so. A new leadership had emerged, and crucial challenges of the community were being resolved. A community-based platform, the Ekta Yuvak Mandal was solving community issues. A sense of despair had turned to hope at the individual, household, and community level. **Our learning was that a game of volleyball could go a long way in motivating youth to transform their communities.**

Community Development in Guptanagar, Ahmedabad

Our second tryst with community development started in 1991 with the communities of Guptanagar, Ahmedabad. Guptanagar is located near the APMC market in the Vasna area of Ahmedabad. A large portion of land on which Guptanagar was located fell under the Urban Land Ceiling Act and hence, to be surrendered to the government. The owners had instead informally sold it to informal land dealers who had then made smaller parcels and sold it to end-users. It was a classic example of "grey market" affordable housing for migrants and displaced people. Since it was deemed illegal by the authorities, there was an absence of basic services of water supply, drainage, sanitation, and waste collection. The residents comprised migrants from Saurashtra region of Gujarat and from the neighbouring states of Rajasthan and Maharashtra. There were distinct communities of Saragra Marwadis, Vankars, Dalits, Vanjaras, Rabaris, Bharwads, and Muslims. When we started working in Guptanagar, there were about 1,200 families.

We met the traditional community, political, and economic leaders of Guptanagar and proposed a partnership for the Integrated Slum Development programme. The leaders welcomed us with the proviso that we should keep them informed.

After promoting male leaders in SMC, we decided to promote women leaders in the community development processes. Worldwide experience has shown, that since women bear the burden of having to take care of their families under all kinds of situations, they become powerful, sensitive, and effective leaders and, have changed gender stereotype images.

We expanded the Saath team to include women. They interacted with women from all communities, on their concerns and difficulties. Their findings revealed their major issues were related to health and children's education. We planned Community Health and Non Formal Education programmes.[4]

During our general women's meetings, we had provisionally identified those who had shown enthusiasm and a desire to bring about change. Six of these women were recruited. Apart from potential leadership qualities, these women could read and write, and belonged to different communities.

They were subsequently trained as community health workers and teachers. A significant part of their work was to visit households and meet parents and patients. Regular reflection of these visits led to a deeper understanding of the causes of ill health and school dropouts. A detailed socio-economic survey of Guptanagar was carried out and analysed; the health workers and teachers were included in the process. The survey indicated that apart from health and education, lack of basic services, livelihood opportunities, and financial services were the other main concerns of the residents.

By 1994, a group of twelve women from different communities of Guptanagar were working as health workers, teachers, and tailoring instructors in various ISD programmes. They had gained credibility and visibility as effective women managers and representatives of the different communities. Historically, the communities had never interacted with each other. However, working together, the women realized the trials they were confronted with, were similar. They had overcome the traditional hostility and suspicion, which was one the causes for lack of development in Guptanagar.

In 1995, this group promoted a women's organization – Sakhi Mahila Mandal (SMM). Before forming the organization, we had

---

[4] Described in Chapters 8 & 9, Pages 93 & 102 respectively

advised them to hold a consultation with the male leaders. They questioned the suggestion. We explained that it was a strategic approach, as otherwise the men would feel threatened and oppose the formation of the Sakhi Mahila Mandal. It would give the Sakhi Mahila Mandal some breathing space to organize themselves more effectively to deal with any issues that could arise from that quarter. Later, a confrontation did take place.

In 1996, through the Slum Networking Project (SNP)[5], the process of providing basic services started in Guptanagar. The Sakhi Mahila Mandal members understood the change that SNP would bring to their neighbourhood, and became intensely involved in the planning and execution. By then, they had become self-confident and effective managers of the education, health, and skill training programmes. They were trusted by all communities and recognized as leaders. They worked together towards a common goal. For the SNP, members of SMM convinced and cajoled their husbands to pay Rs. 2,000, the household contribution, as per the project plan. They consulted households on the layout plan and collected the monthly installments. When SNP was faced with a major hurdle and the infrastructure work was delayed, they convinced skeptic residents to be patient. During implementation, they argued with contractors and AMC engineers to ensure quality infrastructure work. They promoted a women's savings and credit cooperative, Sakhi Bachat Mandal, which enabled households to save up and pay their component of Rs. 2,000 for the SNP. We know that without the women of the Sakhi Mahila Mandal, SNP would have remained a distant dream.

By 2001, when SNP was completed and the health, education, financial inclusion, and skills training programme were showing a positive impact, Guptanagar was transformed from a typical slum to a dynamic and cohesive neighbourhood, especially after residents started upgrading their houses. The women of Sakhi Mahila Mandal were largely responsible for this positive change.

---

[5] Described in Chapter 3, Page 26

Then the confrontation between Sakhi Mahila Mandal and the chauvinistic men of Guptanagar took place. The Sakhi Mahila Mandal took its members for an annual picnic every year. In 2001, SMM decided to go for an overnight picnic. A few men, reputed chauvinists, objected, saying they should accompany the women to protect them. This was not acceptable to the women. On the evening the women were scheduled to depart, the troublemakers stoned the buses in which they were to travel. We decided not to intervene, as we wanted the women of SMM to resolve the conflict independently, which they did, in an impressive, mature way. The women first talked to their husbands and asked them to convince the troublemakers that they, the husbands, did not mind their wives going overnight, so who were they to object. The husbands intervened and firmly told the troublemakers to desist. Then the troublemakers pleaded that two of them should be allowed to accompany the women. The women maturely acceded to this request. Later on, the women claimed that they made the men who accompanied them do all the hard work of lifting and carrying all the groceries and equipment they had for cooking at the picnic, as well as cleaning the utensils. This incident demonstrated how an enlightened leadership, led by women, can resolve community conflicts without external intervention.

Community development in Guptanagar was validated by the emergence of effective women's leadership. The breaking of inter-community barriers, participation by way of decision-making, financial investment, and implementation of development activities, dialogue with traditional leadership in a fruitful way, and engagement with decision-makers to work towards a common good.

Our third community development initiative was more complex

In February 2002, Ahmedabad witnessed major riots between Hindu and Muslim communities. Many Muslim neighbourhoods were devastated and burnt. We decided to implement the Integrated Slum Development (ISD) programme in the neighbourhoods of Saraspur,

Ram-Rahim Tekra, Juhapura and Berhampura. Though Juhapura was not physically attacked, it hosted families who had fled from affected areas in Ahmedabad and other parts of Gujarat.

Members of the Sakhi Mahila Mandal (SMM) and the Ekta Yuvak Mandal (EYM), mainly from Hindu communities, led the ISD efforts. The challenge for them was to work with Muslim communities in affected areas. Initially, there was considerable resistance from the Muslim communities, who were justifiably suspicious. Patiently, the Hindu women and youth of SMM and EYM cultivated a good relationship with the Muslim women and youth, through meetings at the relief camps. Gradually trust was built and ISD programmes were designed and implemented to rehabilitate the affected slums. Women and youth in the affected slums, were mentored by the SMM and EYM members to manage ISD programmes. These women and youth became leaders and formed local organisations. Sankalp Mitra Mandal and Sankalp Bachat Mandal were formed in Juhapura. They were actively involved in the Slum Electrification Programme[6]. They negotiated with the Ahmedabad Urban Development Authority to provide basic services in Juhapura.

In Saraspur, Fatehwadi and Berhampura, women started saving with the Saath Cooperative. Men restored their livelihoods by getting affordable credit from the Cooperative.

We call this process complex, because in addition to the standard community development outcomes of creating an enlightened leadership and a better quality of life, these processes brought about long term communal harmony. When the men and women from the Hindu and Muslim communities worked together, they realized that they faced similar challenges as vulnerable low-income communities and, the perceived myths about each other, were just that – myths, as human beings they had the same aspirations and fears.

---

[6] Described in Chapter 4, Page 46

CHAPTER 8

BRINGING WOMEN TO THE FORE AND PREVENTIVE HEALTH

Women's empowerment has always been at the core of Saath's work. The role of women is central to the development of their families, communities and society; she bears, nurtures and educates her children. Many a time, she shares the responsibility of earning a livelihood. Women living in low income neighbourhoods, more often than not, have to collect and store water, procure fuel for cooking, maintain a household, within a very limited budget, in an unhygienic and crowded environment. She does this in a cultural milieu where women are generally considered inferior and do not have a say in major decisions regarding her family, as well as herself. Empowerment positively influences all round development initiatives – for her family, and for herself.

Our work towards empowerment of women has encompassed livelihoods, health, education, financial inclusion and creating collectives of women.

Preventive Health Care for Women and Children

Public health systems in India do not effectively cater to the health needs of the urban poor. An excerpt from "The Urban Health Context – A Situation Analysis":[7]

"Despite the supposed proximity of the urban poor to urban health facilities their access to them is severely restricted. This is on account of their being "crowded out" because of the inadequacy of the urban public health delivery system. Ineffective outreach and

---

[7] http://nhm.gov.in/images/pdf/NUHM/Implementation_Framework_NUHM.pdf

weak referral system also limit the access of urban poor to health care services. Social exclusion and lack of information and assistance at the secondary and tertiary hospitals makes them unfamiliar to the modern environment of hospitals, thus restricting their access. The lack of economic resources inhibits/ restricts their access to the available private facilities. Further, the lack of standards and norms for the urban health delivery system when contrasted with the rural network makes the urban poor more vulnerable and worse off than their rural counterpart."

"The urban poor suffer from poor health status. As per NFHS III (2005–06) data under 5 Mortality Rate (U5MR) among the urban poor at 72.7, is significantly higher than the urban average of 51.9, More than 46 percent of urban poor children are underweight and almost 60 percent of urban poor children miss total immunization before completing 1 year. Poor environmental condition in the slums along with high population density makes them vulnerable to lung diseases like Asthma, Tuberculosis (TB) etc. Slums also have a high-incidence of vector borne diseases (VBDs) and cases of malaria among the urban poor are twice as high as other urbanites."

We were made aware of this situation during our socio-economic surveys in Sankalchand Mukhi ni Chali (SMC) in 1989 and, Guptanagar in 1991. The survey clearly indicated that stagnant water due to a lack of drainage infrastructure and poor sanitation lead to high incidence of water borne diseases. The crowded and polluted environments lead to respiratory illnesses, including tuberculosis. Children, who were not immunized were vulnerable and fell sick more often. Pregnant mothers had little or no understanding of personal care, nutritious diets, and delivered children at home. They were not aware of the immunization schedule. A general lack of awareness of preventive health care exacerbated the situation. Expenditure on curative health was a significant burden in low-income neighbourhoods. Further, largely due to the unresponsive public health system, those who could afford private hospitals did not use the public health system.

Mother and Child Care in Guptanagar

Our first initiative in Guptanagar in 1991 was a comprehensive Mother and Child Care (MCC) programme. The first objective was to create awareness in the community, especially among women, on safe pregnancy, immunization schedules during and after delivery, affordable and nutritious diets and benefits of delivering children in hospitals. The second was to create a cadre of Community Health Workers (CHWs) and, the third was to establish a local dispensary, which, apart from providing MCC services would become a safe nodal point for women to discuss pregnancy related anxieties, myths, and all-round care.

We recruited a female doctor and enrolled four women from different communities, from Guptanagar as CHWs. The selection criteria stipulated that they should be mothers themselves, have basic literacy skills, be effective communicators and willing to take initiatives. They received intensive training from Sanchetna, an Ahmedabad based NGO with expertise in community health.

The Preventive Health module included identifying the most common diseases and how they spread, their prevention through personal, household, and environmental hygiene techniques and, the importance of immunization. They were also trained in identifying common illnesses in adults and basic treatment at home. The Maternal and Child Care module taught the CHWs to help women have healthy pregnancies and safe childbirth, caring for young children, and family planning. The CHWs learnt communication skills, which enabled them to persuade women and children to question harmful traditional myths and, accept and adopt new methods.

The CHWs then mapped all the 1,200 households in Guptanagar and identified pregnant mothers and new mothers with children between the ages of zero to six years. This long and arduous task enabled the CHWs to get to know each of the eight different communities living in Guptanagar and, understand their different beliefs and myths regarding care during pregnancy and delivery, care of the new-born, immunization, and common illnesses, as well as existing health care providers and the local public health systems. With this knowledge, we created a strategy for our Community Health Programme.

The strategy consisted of providing health cards to pregnant women and children below the age of six. To facilitate regular immunizations, we tied up with the municipal public health department. We promoted the importance of nutritious diets and safe deliveries in hospitals.

Gradually, the CHWs gained confidence and were able to deal with prevailing health issues. Regular immunization camps were held in each community cluster. The community's confidence in the doctor and CHWs increased. Women were able to discuss health, pre and post-natal care, and family planning issues during their visits to the dispensary. Women soon realized that though they were from different socio-economic backgrounds, the challenges they faced were similar. The prevalent myths and superstitions were gradually being put to rest. They started concentrating on their need for better sanitation and water facilities, educating their children, and ways of enhancing livelihoods.

We were able to initiate a nutrition support programme for pregnant and lactating women. A food support package consisting of moong dal, milk and eggs was made available at a subsidized cost. This not only created awareness about affordable nutritious diets, but also reduced harmful dietary practices. The distribution was decentralized – five community level centres were located in the houses of CHWs and participating women. During the weekly distribution, women discussed their experiences and benefits of their nutritious diet. Slowly but surely, we observed their diets changing for the better.

Between 1991 and 2005, 7,300 children were immunized and 800 women participated in the Mother and Child Care program in Guptanagar and Behrampura.

In 2005, the community health programme was merged with the Ahmedabad Municipal Corporation's Reproductive and Child Health (RCH) programme. Some of the CHWs were recruited in the RCH programme. We were asked to coordinate the activities of the RCH in the Paldi and Vasna wards comprising almost 50,000 households and 27 CHWs.

We actively participated in the National Pulse Polio campaign. From 1995 to 2000, CHWs, teachers and volunteers from Saath ensured that all children between the ages of 0 – 5 years were given Polio drops in the Vasna and Behrampura wards of Ahmedabad on National Polio Day.

Checking Tuberculosis

The prevalence of tuberculosis (TB) is quite high in slums. Overcrowded housing and lack of ventilation, along with a lack of water and drainage facilities provide an ideal breeding ground for tuberculosis germs. Our socio-economic surveys of Sankalchand ni Chali in 1989 and Guptanagar in 1991 clearly showed approximately 1,200 TB patients.

The treatment of tuberculosis is complex; the patient has to take a combination of drugs over a period ranging from three to nine months. The drugs are quite strong and if not accompanied by a good diet, can weaken the patient. The symptoms start reducing after a month or so of regular treatment and the patient starts feeling better. But the treatment has to continue till x-rays and blood tests indicate that the TB bacterium no longer exists in the body.

In 1989, we started our TB control initiatives in Sankalchand ni Chali, which reportedly had the highest incidence of TB in Ahmedabad. When we talked to the TB patients, they complained that the doctors at the Ahmedabad Municipal Corporation's TB hospital, which was located less than a kilometer away, were not giving them medicines regularly and, that they could not afford to buy them from chemists. We then met the doctors at the AMC TB clinic. They informed us that they had stopped medication to the patients from SMC because they did not take them regularly, leading to repeated and more resistant recurrences.

After investigating the matter, we realized that both the patients and doctors had reasons to be aggrieved. The perspective of the patients was that strong TB medicines made them weak and unable to work. As they were mostly daily wage earners with limited income, they wanted to resume work as soon as possible. Therefore, they stopped taking medicines when they started feeling symptomatically better.

The doctor's perspective was that patients were not serious about the treatment and, this resulted in more potent TB relapses, which were difficult to cure.

We, along with the newly formed youth group in SMC, developed a strategy for treatment and cure of TB patients. The patients were instructed on the necessity of a long duration treatment. They ensured that patients completed the course of treatment, and had a nutritious diet, which enabled them to continue to go to work. We assured the doctors that we, and the youth group, would ensure that the patients completed the treatment. On that assurance, the doctors agreed to treat the patients we recommended.

Existing and potential TB patients in SMC were taken to the TB hospital for diagnostic tests. Those found infected, started treatment. We then started providing a nutritive complement consisting of moong dal, eggs, and milk on a weekly basis to all the patients under treatment. However, this was given to patients who ingested their medicines in the presence of a youth group member. The youth group members also spread awareness on precautions to be taken to prevent contamination. The youth workers kept a record of the patients. **The strategy worked, and we soon observed that patients were getting cured and the number of new patients was decreasing.** The doctors were happy and started referring more patients to us.

Little were we aware that the strategy adopted by us in SMC in 1989, was also being considered as a viable option by public health institutions for treating TB. In 1995, the World Health Organisation[8] adopted the Direct Observation Treatment Strategy (DOTS) for the treatment of TB worldwide. Unknowingly, we had become pioneers!

In 1996, we merged our TB programme with the newly formed DOTS intervention of the government and started running DOTS centres in Vasna and Santoshnagar. Till date, we have identified and treated 1,180 TB patients in Behrampura, Vasna, and Juhapura.

---

[8] http://www.who.int/tb/publications/manual_for_participants_pp51_98.pdf
https://www.tbfacts.org/dots-tb/

Jeevan Daan

In 2005, in partnership with Counterpart International, a US-based organization, and the Ahmedabad Municipal Corporation, we scaled up our mother and child-care programme through the Jeevan Daan Maternal and Child Survival programme. This was supported by USAID. This programme was carried out in slums in ten wards of eastern Ahmedabad, covering a population of 308,500. Approximately 108,817 children below the age of five and, 19,519 women in the reproductive age group participated in this program. The rationale for the program was the high morbidity and mortality rates among women and children in slums, in stark contrast to the growing economic performance of Ahmedabad city.

Poor neonatal care, premature births, low birth weight, malnutrition, diarrhea, pneumonia, and tetanus were the major contributing causes to the mortality among the under five year olds. Poor access to health care exacerbated the situation.

The under-five mortality rate for Gujarat State was 85 percent, the infant mortality rate of Ahmedabad city was 76. 28 percent of all child mortality was due to diarrhea; Acute Respiration Infection (ARI) accounted for 22 percent and 50 percent were due to poor sanitation and lack of access to clean drinking water.

The baseline survey revealed that within the additional program area, there was an extensive need for health education to improve home case management. Just 15.9 percent of mothers knew how to prepare ORS correctly; 9 percent mothers could cite fast or difficult breathing as a danger sign of Pneumonia. The survey also revealed that in the new slum areas to be covered, 52.3 percent of children between zero to five months were exclusively breastfed, and 24.7 percent were fed within the first hour of delivery. It showed that 22.7 percent of children, zero to twenty-three months, were moderately malnourished. Thirty-two percent of vaccination cardholders were fully immunized.

The main goals of the programme were, sustainable reduction of maternal, newborn, and infant mortality and morbidity, and capacity strengthening of the local partners, Saath and Ahmedabad Municipal

Corporation, to carry out sustainable maternal and child survival activities.

What Jeevan Daan aimed to achieve

The main objective of Jeevan Daan was to improve maternal and child health and prevent, recognize and manage diarrhea and pneumonia. This was done by increasing the capabilities of the caregiver, family and communities through knowledge and positive health practices. The other objective was to improve the quality and accessibility of services provided by AMC personnel and AMC health facilities/services and, establish critical linkages among community organizations, and public and private health providers.

We expanded our team to include a doctor and 12 community health workers to implement the program. We first carried out a Knowledge, Practice, and Coverage (KPC) survey which gathered information on the communities, the prevailing situation with regard to pre and post-natal care. This information was used to build a consensus among project partners, to prioritize and strategize programme deliverables, to develop communication and education messages based on mother's knowledge and practices, and as a reference point for measuring the impact of the programme.

A Health Facility Assessment (HFA) was carried out on the quality of care at AMCs outpatient health facilities. This included staffing, clinic organization, equipment requirements, drug and material supplies, and case-management practices, as well as the type of pre and in-service training of outpatient health workers, and MIS systems for monitoring performance of staff.

We did qualitative assessments to supplement the quantitative data from the KPC survey to provide an understanding of the lives of urban communities, their challenges, positive deviant practices, myths, beliefs, and awareness about health issues. The methods and tools used were Focus Group Discussions and interviews with community leaders and health workers; mapping for seasonal trends of childhood illnesses and the severity among children under the age of five; understanding the community's predilection for health services for maternal and child health, and factors that act as depressors for the low rate of institutional deliveries and, a Doer- Non-doer Analysis to assess the key reasons affecting various behaviours and practices.

The trove of qualitative data gathered was complied through large and small group activities under the categories of: Key issues/challenges/difficulties; Current Beliefs and practices; Knowledge levels; Decision making; Preferred health services; Felt needs of the community; Suggestions from the community for each intervention; Special note/quotes of community member or health staff and Community best practices and resources.

The knowledge and information from the KPC survey, health facility assessment, and the qualitative data compilation was used to make a detailed programme implementation plan (DPI). This was carried out in consultation with AMC health managers, a technical advisory group, health workers, community members, and the programme staff.

Implementation of Jeevan Daan

The key elements of the implementation strategy were, community participation through Community Health Teams (CHTs), Behaviour Change, and Communication Activities (BCC), Capacity Strengthening of AMC out-patient delivery service and establishing a robust Health Management and Information System. The five key program deliverables were:

1. Maternal and Newborn Care: Prenatal visits, Coverage of IFA Tablets and TT vaccination, Place of delivery and Delivery practices, and Post partum visits.
2. Control of Diarrheal Disease: Prevalence of Diarrhoea, Use of Oral Re-hydration Therapy during a diarrheal episode, Knowledge of correct method of preparing ORS, Diarrheal prevention through appropriate hand washing practice.
3. Pneumonia Case Management: Seeking immediate care for a child showing signs of pneumonia, Knowledge of key pneumonia danger signs and of danger signs of a childhood illness (IMCI).

4. Nutrition/breastfeeding: Breastfeeding practice. Early initiation within the first hour of delivery. Exclusive breastfeeding up to 5 months of age. Introduction of complementary food. Prevalence of malnutrition.
5. Immunization: Children 12–23 months fully immunized, including the measles vaccine.

Community Health Teams (CHTs)

Community Health Teams formed the core of the Jeevan Daan Programme. The programme envisaged the participation of communities in decision making and service delivery of the key program interventions and ensuring sustainability after the program was completed. Members of the CHTs were mostly women between the ages of sixteen and sixty, from the slum areas – adolescent girls, young women, newly-wed mothers, mothers, mothers in law who showed enthusiasm, concern and leadership during the KPC and qualitative survey. The size of a CHT varied from four to twelve members, depending on the number of households in a slum, social diversity, and the number of children under five years. Sixty-seven CHTs were formed with around 520 members during the project.

Members of the CHT were made aware of the program interventions. Their skills for communication, record keeping, management and counseling were enhanced. The main functions of the CHTs were to create awareness in the communities to enhance the health status during pregnancy and lactation, form linkages with municipal and private outpatient clinics, create a demand for quality services, and maintain community data relating to the incidence of illnesses, births, deaths, pregnancies, next due date for children's vaccination, etc. They visited private practitioners regularly, apprising them of WHO protocols, and subsequent monitoring. CHT members issued referral chits to increase access to AMC health centres, as well as deal with the bureaucratic procedures for getting sanctions for garbage disposal and water supply. CHTs formed linkages with Aanganwadi workers, teachers, health workers, private doctors, key community leaders and NGOs to improve access to nutrition, immunization, counseling and medical supplies such as ORS, Iron, and Folic Acid tablets.

CHTs interacted with communities through household and community meetings, street plays, visual media, and role plays, using written and audio-visual communication material developed for the program. They encouraged women, children, and youth to participate in these activities. CHTs got together at annual events to realize their collective strength, share experiences and develop newer strategies.

How changing behaviours is key to a stronger community

In Jeevan Daan, a key component to promote positive maternal and child health behaviours at the household and community levels was, through its innovative BCC strategies. The BCC strategy was based on infotainment as a medium and the local epidemiological calendar as a tool. It utilized urban specific, gender and culturally sensitive activities, which included a mix of multiple mediums, including individual and family counseling, street-plays, posters, flip books, exhibitions, games, as well as the traditional cultural mediums. Youth groups and community volunteers with CHT members conducted these activities after appropriate training. A particularly effective method was establishing community kitchens where the women themselves prepared nutritious meals suitable during pregnancy and lactation. BCC activities had a significant impact and were sought by Aanganwadi workers and other health related programs of the AMC, as an effective communication tool.

Another significant achievement of the Jeevan Daan programme was a community-based health management system run by CHTs, which enabled appropriate preventive and curative health measures.

Model Urban Health Centre

In 2005, we partnered with IIM Ahmedabad, Gujarat Cancer Society, and the Akhand Jyot Foundation, to design and establish a model Urban Health Centre with focus on people living in slums and chawls. The model was based on extensive use of Geographic Information System (GIS) to locate urban health centres in the Vasna, Sarkhej, and Paldi wards to ensure availability, access, affordability, and equity of healthcare services, as a viable alternative to private health care providers.

Our team carried out extensive household surveys in these wards to assess the health needs of low, middle, and high income groups and their current practices of seeking health care services.

The UHC which was located in the Vasna area was the first of its kind in India providing a comprehensive line of services under one roof, including consultation, lab and radiology services, medication, and referral services. This reduced the cost of transportation and time taken to give samples and collect reports.

The residents were able to avail quality health services at affordable rates and not be at the mercy of private doctors/quacks who preferred to prolong treatment and use more than the required medicines.

CHAPTER 9

CHILDREN AND EDUCATION

The girl with grit

The 2002 riots in Ahmedabad had far reaching consequences. Sanjida Ahmed Shaikh, from Juhapura, was one of its victims. Ever since she had lost her husband to the riots, she had been in a terrible emotional state. She was depressed and refused to go anywhere or talk to anyone. One day, her daughter cajoled her to make an effort and took her to school where a meeting was in session. There she was asked to take down minutes of the meeting as she was educated and could write. "It was my point of no return," she smiled.

It was during this time that Saath had made inroads into Juhapura. Members from Saath, especially, Chinmayiben, Devuben, and Kailashben encouraged her to join the Balghar – a preschool – as a teacher. She went through trainings and became a teacher at the Balghar in 2003. She was also inducted into the ways of dealing with community resistance with patience and confidence.

Later, she came to meet us, proudly introducing two of her students from the 2005 batch. Rukhsana had been a pale, sickly child and her parents feared that she had Tuberculosis. But a checkup cleared the doubt. With supplements she improved. Her mother was advised on nutritive foods for her children. Rukshana 's mother was illiterate and didn't want her daughter going to school, when she could learn household chores. At the end of every school year, Rukhsana's would tell her she had to give up school and stay at home. But Rukhsana was determined to finish school. With constant encouragement from Aasma, she persisted, and soon her mother too began to see the wisdom of education. After completing her schooling, the family shifted out of Juhapura to Vatva. Rukshana didn't feel very safe or comfortable in her new environment. While studying she had done a course in Basic English as well as computer training through the Saath Centre. She now works for a motor company; started out in the Service Department but is now in the Insurance Department. She feels her parents put a lot of trust in her as she travels a very long distance to her place of work.

"I have had a varied career up till now; I worked for a phone company and then as an executive agent for a company selling cigarettes online," she said with a huge smile. She went on to explain that the cigarette company was the number two in the world and, she made very good sales numbers.

She has completed the Computer Teacher's Training course and is awaiting the result. She is also hoping to get an international certificate soon. In a wonderful twist of events, today, Rukhsana's mother is also a teacher at a Balghar.

India is home to the largest number of children in the world. According to the "Status of children in Urban India," conducted by the National Institute of Urban Affairs, New Delhi and based on the 2011 census, 472 million children, who are 19 per cent of the world's children, live in India. Every fifth child in the world lives in India.

Urban India has 128.5 million children. 36.6 million children are in the 0–5 years age group. 26.4 million children are in the 6–9 years age group, 35.9 million are in the 10–14 years age group and 29.6 million are in the 15–18 years age group.

65.5 million people in urban India live in slums and constitute 17.4 per cent of the urban population. Out of this 8.1 million (12.3 percent of total slum population) are children in the 0–6 years age group. The census of India does not enumerate population of children above 6 years of age. Another 0.11 million children live on the streets. Various studies indicate low learning levels and a high dropout rate among slum children.

The challenges for a decent education for children living in slums are many, and daunting. The milieu for learning is almost absent. Most children are first generation learners, as their parents have either never gone to school, or, may have dropped out in the early grades. Parents have little understanding of the requirements of the present-day education system in schools, in terms of teaching aids, text-books, extra-curricular activities and even less on creating a learning atmosphere at home. In most households, both parents work as skilled or unskilled labourers, they are unable to pay adequate attention to their children's education. The physical environment is an impediment as well. The congested houses leave little space for children to study comfortably at home. The unsanitary conditions and high levels of pollution, as well as a shortage of water, lead to a higher incidence of sickness and absenteeism from school. Children from slums are generally not "properly" clothed. These parameters, when compared to those of other children attending school, lead to a condescending attitude from teachers who label them – "underachievers."

A majority of children go to municipal schools where the level of teaching is not on par with other schools. A combination of these factors is a major drawback in terms of being able to cope with their studies. Various studies have indicated that learning levels among children from slums are at least two grades below their counterparts from middle-class families.

Like most parents anywhere in the world, these parents too have high expectations of their children's abilities, specially pertaining to their education. They do their best, within their limited capability and understanding, to get their children a good education. They send their children to private schools or tuition classes. In spite of their best efforts, they find that their children do not do well and ultimately drop out of school, either at the 7th or 10th grade. Girls fare far worse, as they get inadequate attention due to traditional attitudes, which include the burden of helping with household chores, or looking after younger siblings.

The long-term effects of school dropouts are detrimental not only to the children, but their parents as well. Children start believing that they are not capable of learning, have low self-confidence and end up doing menial jobs. Parents, who have invested much in their children's education, start believing that their children are inherently inferior and their investment in their education is a not worthwhile. It is a vicious circle sometimes, even for those who have a college degree, when they are unable to get a job and have to start learning a skill, very often the same one their parents have.

In 1993, after analysing this situation, we decided to focus on building a strong foundation at the pre-school level, as well as supplementary education for the children residing in Guptanagar, to enable them to cope with school. To encourage a supportive learning environment at home, we encouraged parents to get involved in all aspects of their children's education. We started preschool classes called Balghars, and supplementary classes for school-going children.

What are Balghars and why they work?

Balghars are pre-school classes for children between the ages of three and five. They are run in rented rooms within the slum, where parents and communities can observe the activities and engage with the children and their teachers. Balghar rooms are designed to facilitate formal and incidental learning. Typically, a Balghar has one room and a courtyard where children can sit, run, jump, climb, play games and also learn through group activities. There are attractive charts on the walls for incidental learning. Various teaching and learning material, games, and toys are included. Typically, a Balghar is run by two teachers, for about thirty children.

The teaching methodology in Balghars is based on Jean Piaget's theory of cognitive development, a process by which children construct an understanding of the world around them, then experience discrepancies between what they already know and what they discover in their environment. The emphasis is on overall growth of the child by paying attention to social, mental, physical, language and maths, creativity, and environment awareness and development.

Who are the Balghar teachers?

The teachers in Balghars are committed women, carefully selected from within the communities. These women do not have a formal education in teaching. A majority of them have studied till the 12th grade. They are women who are interested, show initiative, and are willing to learn. They demonstrate a willingness to work with children, understand the need for education as a means of developing their families and communities. They undergo intensive training in child education, development, psychology, and experimenting with new methods based on Piaget's model. They learn the methodology of observation, innovative teaching and case studies. They prepare files on the progress of each child in their classroom. Teachers independently manage all aspects of the Balghar programme.

The role of parents in Balghars

The parents are the most important stakeholders in the Balghar programme. The programme is designed so that parents are continuously engaged with both, the education of the child and the running of the Balghars. Teachers visit the children's parents once a month to discuss the progress of the child. They take with them the individual file of the child, which contains a record of every activity carried out by the child. Parents are pleasantly surprised when they see the work done by their children. They are also encouraged to talk about the new activities and learnings of the child outside the classroom. Parents are encouraged to visit the Balghars and observe their children. Teachers plan the menu and help in preparation of the daily meal.

Parents appreciate the Balghar programme because their children have become interested in learning instead of just wandering around. They have observed that their children had become punctual and regular in studies; can now talk, sing, and behave in a more acceptable manner, and are clever and independent. Parents appreciate the organized visits and celebrations, as they are unable to provide that kind of exposure to their children.

The community as a whole are instrumental in enhancing the confidence of the parents and teachers in the Balghar programme, by actively being involved in the process.

Monetary contributions through fees are an integral part of the Balghar programme. The monthly fee per child has increased from Rs. 20 in 1993 to Rs. 100 today. The fees cover the cost of meals provided to the children, but not other operating costs. Paying fees gives the parents a feeling of entitlement, which leads to checking on the progress of their children. The teachers become accountable to the parents and the community. We have seen that almost all parents pay the fees. Concessions are given to the few that cannot afford. **We have observed that children who attended Balghars regularly have performed well in formal schools, and become self-confident adolescents.**

Over the years, we have run 17 Balghars with 3,400 children in Guptanagar, Sankalitnagar with 48 teachers.

Supplementary classes

Children from slums lag behind their non-slum counterparts in primary schools. By the time they reach grade 4, slum-children are, on an average, two years behind the formal curriculum. This handicap continues throughout their schooling, resulting in a high failure and dropout rate.

The supplementary classes, through its innovative and relevant teaching methods help children in the primary school cope with their curriculum. It is not a tuition class. In these classes, limitations of students, in terms of cognitive and psychological needs are identified. Through innovative teaching methods and psychological support, children's self-confidence and self-esteem are enhanced leading to an ability to cope in and outside school.

Similarly, special classes with specifically designed curriculum for non-school going children and dropouts are also conducted. Some of these children then join the formal system. A special effort is made to encourage girls to join these special classes. Adults learn reading and writing skills in literacy classes. Teachers from the community, as well as volunteers with the appropriate skills, conduct these classes. These classes too are not free; regular fees are charged.

Integrated Child Development Scheme (ICDS)

The impact of the innovative Balghars on the communities spurred us on to scale up the Balghars by integrating them with the Integrated Child Development Scheme (ICDS) of the Government of India.

Launched in 1975, the ICDS is a flagship programme of the Government of India for early childhood care and development. It is designed as a response to the challenge of providing pre-school non-formal education on one hand, and breaking the vicious cycle of malnutrition, morbidity, reduced learning capacity and mortality on the other. The beneficiaries under the ICDS include children in the age group of 0–6 years, pregnant women and lactating mothers.

Similar to the Balghars, the ICDS offers a package of six services. These are – supplementary education, pre-school non-formal education, health and nutrition education, immunisations, health check-ups, and referral services.

Though well-funded and designed, the ICDS has certain drawbacks in implementation. The key person for implementation is the Aanganwadi Worker (AWW), who is typically a woman from the community where the Aanganwadi is located. The AWW is overloaded with tasks. Apart from her regular work of mobilizing children, teaching them, preparing their meals and ensuring immunizations, her duties extend to other public interventions, such as TB and AIDS awareness, Polio days, other activities related to the health department, as well as activities related to the education department, such as total literacy programme, Sarva Shikshan Abhiyan and the District Primary Education Programme. The AW is also inadequately trained, paid a low salary, irregularly, and is hampered by lack of teaching aids.

We approached the Women and Child Development Department in Gandhinagar and requested them to allot Aanganwadis in Ahmedabad to Saath. In 2003, Saath was sanctioned 66 Aanganwadis and in 2004 an additional 120 Aanganwadis across 21 wards of Ahmedabad city reaching out to 7,500 children. We soon found that the functioning of the Aanganwadis considerably improved with the introduction of the Balghar methodology. The key element, which led to the improvement was involving parents and the community in the operational aspects. We started charging fees of Rs. 20 per month to increase accountability and participation. But soon hit a major roadblock. The ICDS rules do not permit charging fees. The services are to be provided free of cost, which results in lack of transparency and accountability. Parents, children, and the community do not have a say in the matter and cannot question any aspect of the programme.

Study on ICDS

To convince the authorities that adopting the Balghar methodology and charging fees would lead to better delivery of the ICDS services, we undertook an eight-month action research study to strengthen the quality of the preschool education component in the ICDS. The objectives of the study were:

- Improve skills and motivation of the AWWs in actively managing the pre-school education (PSE) along with other components of the ICDS and lay emphasis on the managerial role of supervisors
- Ensure community participation in effectively carrying forward and sustaining PSE activity
- Establish effective Planning, monitoring and evaluation systems for PSE components
- Improve the physical environment of the AWs

Study methodology

The methodology was a comparative assessment of 48 AWs (almost 25 percent of the total AWs that Saath runs in the various wards). Of these 48, 24 were selected as 'experimental AWs', where special Balghar-like inputs were provided. The rest were 'control AWs', with routine input, and an initial request to include the PSE component in their schedule. The study was participatory, at all stages, beginning from the preparation of study design, action phase and final assessment. The emphasis was on 'learning by doing'. There was plenty of room to accommodate changes/ modifications during the course of the study, depending on the participants' feedback. At the end of the study, to gauge the impact through various identified indicators, a comparative assessment of both groups of AWs was done.

The study clearly demonstrated that the experimental AWs', where special Balghar-like inputs were provided, performed better. A summary of the conclusions of the study were:

a. Introduction of the Pre-School Education (PSE) component led to improved performance of children and an increase in the attendance hours.
b. With a little innovation and marginal costs, it is possible to convert the dull and drab AWs into attractive and neat centres. The impact of PSE can also be found in the level of cleanliness of the children.
c. With appropriate time management and following set schedules diligently, it is possible to manage PSE along with the other workload. The quality and quantity of the activities do not suffer.
d. Computerization of records and supervisors' working knowledge of computers help in curtailing unnecessary wastage of time in record keeping.
e. A planning, monitoring, and evaluation (PME) system is the key to the success of PSE. In the absence of a well-defined and structured system, it may be difficult to sustain the quality of the programme.
f. Community participation is considered very important for sustaining the PSE. It was demonstrated that it is possible to invest in children's needs such as toys, teaching aids, educational visits, etc. from small individual contributions, thus reducing dependency on the ICDS department and also helped make the

teachers more accountable. Recommended to lay more emphasis on this aspect of community participation.

g. Appropriate and interesting teaching aids played a pivotal role in making an impact in the study.

h. Community involvement such as involvement of adolescent girls/mothers in helping with management of AWs has shown positive results. It has not only helped the AWWs function better, but also resulted in more ownership of the AW on the part of the mothers/young girls. Likewise, regular parent meetings and their involvement in the decision-making processes of the AW helped sustain the programme. Celebrations/visits (through an activity component of the PSE) are seen as a major source of fostering community involvement and acknowledging the efforts of the AWWs.

i. Refresher and hands-on training to the AWWs are a must to improve and hone their skills. The study showed that this not only helped the AWWs of the experimental groups to perform better, but also was responsible in their increased confidence and motivational levels.

We presented the study to the authorities to convince them to allow Saath to run the ICDS centres based on the Balghar methodology. They were however, not convinced and we decided to withdraw from the ICDS programme in 2007. We felt that we could not simply be a government contractor providing sub-standard services, which we had demonstrated, could be of a much better quality.

Child Rights for Change

Between 2009 and 2013, Saath implemented an ambitious Child Rights for Change project in partnership with Save the Children/Bal Raksha Bharat with support from IKEA Social Initiatives. The project's main goal was to prevent child labor in the cotton growing Talukas of Viramgam and Dholka in Ahmedabad district, Gujarat state. A total of 120 villages were covered with a population of about 11,000 people.

The project's rationale was that children from poor and landless families, a majority of whom belonged to the SC and OBC castes, had to work in cotton fields, which not only denied them their basic rights, but also affected their overall development. There were issues relating to child protection such as sexual and physical abuse. To prevent child labour, it was essential to sensitize parents and the farming community at large on the issue, as well as initiate child protection measures and, facilitate an inclusive education process.

The project adopted an integrated approach of sensitizing all stakeholders – children, parents, farmers, village leaders, bureaucrats, school management and teachers, Aanganwadi and Asha workers, women self-help groups and civil society, to the multi-faceted aspects of child labour, as well as working towards bringing about a systematic change.

The strategy consisted of formation of children's groups, village-level child protection committees, energizing women self-help groups, enabling access to rural employment guarantee schemes, strengthening linkages with Aanganwadis, making education attractive and inclusive, and initiating district and state level networking and advocacy.

To this end, a team of 40 Development Activists (DAs), young people from the project villages, supported by a 9-member management group undertook the massive task.

The first task of the DAs was to sensitize all stakeholders and obtain their consent to run the multifaceted project. This was done through general discussions, focus group discussions, plays, skits, and films.

The next task was to form village level Child Protection Committees (CPCs). The members of the CPCs comprised of – the Sarpanch (elected head of the village), a school teacher, an Aanganwadi worker, the village social justice committee chairperson, leader of the Self Help Group (SHG) and 5 community leaders. 50% of the CPC members were women. A project plan for each village was prepared in consultation with the CPC, which met every month to monitor progress and help solve implementation challenges. The agenda of the CPC meetings, which were held at block/taluka level, included awareness and training on Child Rights, Right to Education, ASER reports on school learning levels and testing children on learning levels. CPCs encouraged famers to take a pledge to stop child labour and gave certificates to farmers who complied.

When the Right to Education (RTE) bill was enacted in 2009, CPC members became part of the mandated school management committees. Around 300 orphans were identified and admitted into schools by the CPC. Gradually, CPCs became an institutional mechanism for addressing local child labor issues.

240 Child Groups (CGs) also known as, Bal Panchayats, were formed – two in each in each village. Members of the CGs were children representative of all castes, gender and religion; school going, non-school going, child labourers, disabled, and others – girls and boys. CGs provided a safe platform for children to get together to raise and discuss issues of concern with adults in the community.

Children as primary stakeholders

CGs enabled children to analyse how a particular decision would best serve their interests. By bringing in the children's perspective on issues of child labour, abuse and neglect, it increased the likelihood of the project becoming relevant and sustainable. CGs gave children more ownership and a sense of responsibility. It enhanced their self-esteem and provided an opportunity to learn and practice the skills of responsible and active citizenship. CGs increased the visibility of children's issues and placed them on community agendas, increased adult understanding and appreciation of children's capacities, as well as accountability towards them. CGs maintained a register to track non-school going and irregular attendees. CGs identified teachers who were physically or verbally abusive towards children and approached the school management along with the CPC. 2,200 children were trained in life skills

The key objectives of the project were education and livelihoods of children, with emphasis on female child labour and children of cotton pickers. The programme worked intensively to create awareness amongst parents. A list of pre-schools, irregular and non-school going children was drawn up for the 120 villages along with a list of Aanganwadi centres and schools. Birth certificates were obtained for about 950 children. Almost 290 Aanganwadi workers received training to upgrade their skills on inclusive and effective pre-school learning. A manual of indicators and tools were prepared for assessing learning, reading and writing skills for pre-school and school going children. A child friendly environment was created in 120 Aanganwadis and 120 schools. Vocational and life skills trainings were provided to 320 adolescents. Village Education Committees (VECs) were formed to monitor the education initiatives of the project. They were made aware of the Right to Education (RTE) act of the government. VECs and CECs worked in tandem to ensure good quality education in Aanganwadis and schools, as well as regular attendance.

A major component of the project was to empower women as mothers of children, farm labourers, leaders, and decision makers. The governments of India and Gujarat have promoted the concept of Self-Help Groups (SHGS) to enhance economic and social status of vulnerable women. A typical SHG has ten to twenty members with a homogenous livelihood and social background. SHGs mobilize savings and disburse loans to members, maintain accounts and records, access loans from government schemes and micro-finance institutions. Over time, SHGs have become platforms of empowered women who influence local development decisions regarding education, health, water resources and irrigation and livelihoods.

Through the Child Rights project, Saath enabled below poverty line women to form 240 new SHG groups and strengthen existing groups and provided leadership skills to 3,000 women. By building the capacities of SHGs, women were linked to the National Rural Employment Guarantee Scheme (NREGA) of the government. Eligible women were able get BPL cards to access social benefits.

The farmer community, as a vital stakeholder, was closely involved in all aspects of the programme. They were key members of the CPCs. As CPC members, they encouraged farmers to take a pledge not to employ child labour and certified those who did, at public events. They donated resources for the upkeep of schools and Aanganwadis. In one village, Birani, a farmer donated land for the construction of a school. District level Farmers' Associations were formed to monitor the program and influence farmers, policy makers, and local government functionaries.

Civil Society Organisation (CSOs) from various project districts in Gujarat, Rajasthan, and Maharashtra formed a network for advocacy with Central and State governments to create increased awareness and strengthen the implementation of the Child Labour (Prevention and Rehabilitation) Act (CLPRA). An Action Committee formed by the CSOs prepared plans for strengthening implementation, as well as strategies to make progressive changes to CLPRA. An interstate mechanism for monitoring and coordination of child labour was established for Gujarat and Rajasthan. An interstate civil society network was created to document, rescue, and rehabilitate child labour through government and civil society organisations.

Apart from the significant impact of the project in creating awareness and reducing child labour, the long-term effect has been that the Gujarat Government has mandated the formation of Child Protection Committees in every village. CPCs have the same status as Water and Social Justice Committees and work with the local Panchayats to ensure prevention of child labour.

Upgrading Aanganwadi and School Infrastructure

Saath has upgraded the Aanganwadis and schools with clean drinking water and sanitation facilities, as well as play areas in rural pockets of Ahmedabad and Sanand districts. In consultation with Aanganwadi workers, school management and village residents, Aanganwadis and schools with sub-standard facilities were identified. Water purifier units, rain-water harvesting structures and toilets, separate for boys and girls were installed and play grounds and sports facilities were upgraded and trees planted in 106 schools. Village level committees were formed to oversee construction, installation, and subsequent maintenance.

The impact was significant. More children, especially girls, started attending school. Open defecation decreased and children stayed at school to play and compete in sports events. Ford Motors CSR and the NDTV/Coca-Cola campaign for supporting schools supported these initiatives.

Right to Education Initiatives in Ahmedabad

The path breaking Right to Education (RTE) bill enacted in August 2009, seeks to make education a fundamental right of every child between the ages of six and fourteen and specifies minimum norms in elementary schools. It requires all private schools to reserve 25 percent of seats. Children are admitted to private schools based on economic status or caste-based reservations. It also prohibits all unrecognized schools, donations or capitation fees. It also stipulates no interview of the child or parent with regard to admission. The Act also provides that no child shall be held back, expelled, or required to pass a board examination until the completion of elementary education. There is also a provision for special training of school dropouts to bring them on par with students of the same age. The RTE Act requires surveys that will monitor all neighbourhoods, identify children requiring education, and set up facilities for providing it.

When RTE was introduced, parents of eligible children and school managements were unaware of the procedures required to admit children in schools. In 2012, IIM, Ahmedabad carried out research on the implementation of RTE. Saath and IIMA researchers met parents of children attending the Saath Balghars and school managements to understand their concerns and perceived challenges. We started admission to private schools of children from Balghars, eligible under RTE and realised that parents did not have documents related to income and residence, required for eligibility. With other NGOs in Ahmedabad, a forum was formed to facilitate RTE admissions. The forum met the Collector and apprised him of the non-availability of some documents and the difficulties of admission. The collector formed appropriate guidelines to ease admission.

The forum creates awareness about RTE among parents, facilitates procurement of documents and admissions. It has a list of schools that parents can choose from. The forums mentor students who have been admitted enabling them to cope with their studies, as well as fit into their new environment.

Creating child friendly spaces

The estimated numbers for out-of-school children by various official sources in India, show wide variations. The Ministry of Human Resource Development (MHRD) survey (IMRB-SRI, 2014) estimate 6 million, while for the same year, the National Sample Survey (NSS) is 20 million.

The rate of enrollment of children in schools within the construction community is abysmally low. The lives of construction workers basically are that of nomads, constantly shifting from one construction site to another. They are often unsure of their next location and usually stay at a particular location for less than six months. Moreover, during the monsoons or periods of unemployment, they move back to their native villages. The constant change in location prevents them from enrolling their children in schools. Many of the parents are illiterate and therefore do not value the importance of education. The situation is further worsened by the inadequate facilities for migrating children in most government schools. Lack of awareness about The Right To Education (RTE) Act and its cumbersome procedures, is the major hurdle.

The Child Friendly Spaces program at Saath works with construction workers and real estate developers to educate the children. The curriculum and methodology are similar to the Balghar programme. These classes are conducted at construction sites. Children who are eligible are enrolled in formal schools through the RTE process.

Through CFS, Saath has worked at 14 construction sites, with 7 developers and 5,280 children

Girls Education

Mom's touch

Saath works with NIVEA's CSR programme "Mom's Touch." The programme aims at strengthening families by empowering mothers and providing them a helping hand to ensure their children go to school regularly. The programme supports poor women who work extremely hard to ensure that their children complete their education. The programme provides monthly household groceries to the selected women.

The mothers are selected on the basis of the support they provide to educate their girls. So far 650 children have received support through this programme.

Girl Child Education Program

With Intas Pharmaceuticals Ltd.'s CSR initiative of supporting and enhancing female education, Saath identifies poor and vulnerable children living in areas or villages around the Intas Pharmaceuticals Ltd plants. A scholarship is provided to meet expenses related to school fees, text and note books, stationary, uniforms, shoes, school bags, transportation, lunch, etc. so that they can continue their education.

Saath identifies girls from poor families, orphans, and single-parent families who have demonstrated a good academic record and do not receive any other support. The child's progress is regularly monitored by the Saath team, which also offers counseling support.

So far thirty-two girls have received scholarships amounting to Rs. 5 lakhs in this program.

CHAPTER 10

HARNESSING THE POWER OF THE YOUTH

Anita's story

My name is Anita Mahida and I'm eighteen years old. I was born in Behrampura, Ahmedabad. I live in a family of ten, which include my grandparents, parents, four sisters and a brother. My father used to run a restaurant at the drive-in cinema. He, however, had taken to drinking. His business partner took advantage of the situation and defrauded him and took sole ownership of the restaurant; our only source of income. We were faced with a financial crisis. My father did not recover from this betrayal. The entire responsibility of our family fell on my mother's shoulders.

My mother, Nathiben, is a brave woman; I admire her courage and strength, which she showed during this difficult phase. To make ends meet she started doing household work in nearby areas. All of a sudden, one day, my father abandoned us. Throughout these trying times my mother single-handedly got my four sisters married and also took care of my education. Due to inflation, it was becoming extremely difficult for my mother to provide for the family on her own.

After a long gap of seven years, my father returned. During that period, my brother too had gotten married, but due to our dire financial condition, he went into a depression and committed suicide in 2012. Soon after his death, my father also passed away. It was a double blow to the family. I decided that

I had to lend my mother a helping hand. I quit my studies and joined a call centre. During this difficult time, my mother constantly encouraged me and was a source of real strength and courage. I learned a lot from all the difficulties I faced.

One day, I came across a pamphlet about the Umeed programme at the call centre, I was impressed by the programme and the additional skills that I could learn, like basic computer literacy and personality development. I decided to do a Bed Side Patient Assistant course, as I could continue working at the call centre. I joined the training programme. During the course, I met Sandeep Sir, who was a city coordinator of the Youth Force programme. He told me all about the Youth Force and how it opened up opportunities and gave direction to vulnerable youth to fulfil their aspirations and dreams. I loved the concept of the programme and decided to join the Youth Force.

After joining the Youth Force, there was a definite change in my attitude; I used to be angry and irritated about small things, it changed, I have become calm and mature. Because of my bad attitude, I didn't have many friends, but now I have many, with whom I can share my feelings. The Youth Force has given me the much-needed confidence. Through various social campaigns and activities, it has also made me more responsible towards society. Earlier I was unaware of the social and cultural aspects of the city, but the programme helped me understand them. I would like to thank the programme.

Anita completed her training at the Umeed centre and is currently working at a hospital.

India is poised to benefit from a demographic dividend. Young people aged between ten to thirty-five years form 45 percent of India's population. These young people will shape the future of India by being the harbingers of change. The world over, they are the idealists, full of energy and motivation. Most young people generally dislike being bound by tradition, culture, and history. They yearn for change, yet are often frustrated and helpless, as they feel that they do not have the wherewithal to bring about the change. We find this is more so with young women who are bound by traditional and cultural restrictions.

In slums and chawls, the frustration and dissatisfaction among the youth is more pronounced because the challenges are greater. The degree of disempowerment is higher. Unlike their middle-class counterparts, these young men have little or no education and are less informed due to lack of opportunity and support systems.

They can only afford to attend municipal schools where the quality of teaching is poor. The financial means of their parents are limited, and they cannot afford to buy supplementary educational books and material or attend private tuitions. These factors combine to lead to a high dropout rate in the 10th and 12th grades. The situation among girls is worse, with a high dropout rate in the 7th grade. Consequently, the livelihood options are restricted to entry-level jobs in the formal sector, apprentices in the informal sector, or micro-entrepreneurs by compulsion.

In the language of development, these youth do not have the knowledge, or the economic and political clout, or societal networks to help bring about change; hence their frustration and helplessness. If given opportunities and an appropriate support platform, these enthusiastic young people can bring about transformation at a personal, family, and community level.

Saath has created platforms and support systems for young people to effect sustainable change. These are methods and processes, through which they get together, realize their individual and group potential, enhance their self-esteem and self-confidence, as well as understand and resolve challenges through logical analysis and action. They become progressive and tolerant individuals and leaders with skills and understanding, which enable them to be compassionate individuals, responsible citizens, effective employees, workers, and entrepreneurs.

Youth in Sankalchand Mukhi ni Chali

Saath's engagement with the youth started in Sankalchand Mukhi ni Chali (SMC) in the Berhampura area of Ahmedabad in 1989. SMC was a dilapidated chawl of 500 households with abysmal sanitation conditions, high incidence of tuberculosis and a large number of unemployed youth. Our objective in working in the SMC was to address these challenges by empowering the youth.

During our initial visits and meetings with the SMC residents, there was anxiety on both sides. We were unsure of our welcome, as we were a newly formed NGO of just two people. Furthermore, we came with our arrogant-sounding objective of transforming SMC! The residents were apprehensive; our status being what it was – a new NGO with no track record whatsoever! Fortunately for us, they were trusting, bighearted, and tentatively agreed to work with us.

We reached out to the youth by playing volleyball with them in the evenings. After the games we would talk about the situation of SMC. The youth vented their frustrations on their inability to do much to change the situation. We initiated discussions on how together we could address the issues. What emerged was an action research project; the youth being trained as Social Animators (SAs) would simultaneously tackle the issues in SMC. Our role was that of a guide and facilitator.

Social Animator course and the Action Research project

Playing volleyball regularly. As a sport, volleyball is an inexpensive team game, physically challenging with a time limit. The young men were able to acquire the skills of this new game, which included playing as a team. Gradually they taught the skills they had learnt to other interested young men. They learnt how to strategize and complement strengths and weaknesses towards the common goal of winning. Being part of the team gave them an identity, which was based on skills and merit, unlike the caste and community identities they were used to.

Household Socio-Economic Survey

These young men participated in the design of a household socio-economic survey form. They understood the need for the various types of information that was required and its purpose. When they filled the survey forms, they realized the importance of getting accurate information and recording it correctly. When they interacted with households while filling the survey forms, they were able to build a rapport and more significantly, realized that almost every household faced the same challenges they did.

We collated and analysed the data with them. They learnt and used their mathematical and logical skills. They understood how to apply the information and draw inferences. They learnt how data and its application were closely related to their everyday lives.

Most importantly, they gained more confidence because they had learnt how to design, conduct, and analyse a socio-economic survey – learning through action.

The survey revealed the major issues of SMC were sanitation, tuberculosis, and financial inclusion. The data and analysis provided the information by way of numbers, location, and households. A detailed action plan was prepared to tackle these issues.

Outdoor camps

We accompanied the youth to a three-day overnight camp in the outskirts of Ahmedabad, close to nature. We lived in tents and cooked our own food. They learnt to manage the logistics for conducting the camps; to estimate and procure the rations required and arrange transport. They learnt to pitch tents and set up a kitchen, as well as the various roles and management functions required for running a camp.

During the camps, they participated in various activities, which enhanced their leadership skills. They discovered skills such as public speaking, cooking, negotiating, resolving conflicts, planning, etc. which were dormant within them. Discussion and reflection on diverse aspects of personal, family, and community concerns led to insights on how these concerns could be resolved. A deeper sense of empathy emerged, as they understood that they all had similar challenges. Above all, the camps built a sense of camaraderie amongst them.

The youth met regularly every week to monitor the action plans. This involved discussing successes, challenges, and failures. To solve problems, the required trouble-shooting and strategizing was done. Asking for and giving help, based on skills and networks, began. The need for timely and structured meetings, as well as recording them was recognised.

Formal youth group

They realized that they had the skills and the wherewithal to bring about positive change amongst themselves, their families and community. They understood that they would be more effective if they created a formal platform representing their community and SMC. They created the Ekta Yuvak Mandal and registered it as a CBO (Community Based Organisation).

How did the transformation of those diffident, with low self-esteem, apprehensive, individualistic youth to self-confident, supportive, risk-taking and self-assured young people happen? We believe that it was creating a platform and support system by which they were able to realize their abilities and potential. The action research enabled them to discover their skills and apply them to resolve real life challenges. In the process, they gained in confidence and their self-esteem was enhanced. They started applying the skills and abilities to other aspects of their lives and were able to get better jobs or entitlements. They now had the skills to achieve.

Almost 150 youth were involved in the SA Course and Action Research. As this framework was working with the youth, we applied it in our work with women and children.

Youth Force

Abhishek Maurya is a 22-year-old energetic young man from the Behrampura Slums of Ahmedabad. He belongs to a low middle class family. He had never been out of the city, except for a family trip to Rajasthan when he was 18 years old. Behrampura is where the Slum Networking Project (SNP) of Saath began around two decades ago. At that time, Abhishek's father, a tailor in the locality, started actively participating in the activities of the SNP. Although the family lived in the slum cluster where the Umeed and Micro-finance centres were located, they were not aware of the NGO sector or other career options.

After graduating from school, Abhishek completed his B.com. One day he met Sandeep, the Youth Force City Coordinator, who was in the locality for community mobilization. Sandeep informed him about the Youth Force, which provided a platform for the marginalised urban youth to explore their horizons. Initially, he was hesitant, as he did not want to devote his Sunday for the task. Sandeep, however, kept motivating and encouraging him. Sandeep told him that he could also get a chance to play cricket again, his childhood passion. Joining the Youth Force would enable him to enhance his skills and gain knowledge, which would help him acquire gainful employment. Sandeep's efforts paid off. Abhishek joined the Youth Force, as a member and, also enrolled for the Tally course at the Udaan Centre. Post training, the Saath team offered him the post of City Facilitator in the Youth force.

He started his career with Youth Force with a monthly package of Rs. 8000. His work involved door-to-door campaigning, orienting visitors, organising in-door and outdoor meetings for the youth and taking Life Skills sessions at the Udaan training in different cities of Gujarat. He visited places like Goa, Shirshi Town in Karnataka, Mumbai, and Delhi. He attended the Urban Youth Animator leadership training organised by Yuva, an NGO in Navi Mumbai. He took part in the training on Environment conducted by Earth Watch, in Shirshi, a remote forested hilly terrain on the Western Ghats, which was a great learning experience for him. He learnt about Life cycle changes when any particular species in the eco-system are destroyed and how that affects the output on farmlands. These exposures helped shape his personality. He is a confident young man.

During these years of association, he has learnt to work with grassroot communities and the socially marginalised Youth. He has also been facilitating corporate representatives with campus placements from among the Udaan trainees, carrying out follow-ups and creating linkages for job placements.

"I have been given the opportunity of speaking in English after joining the Youth Force; since I belong to a low middle class background no one had ever motivated me to do so," said Abhishek. "What I love the most is that the Youth Force enables me to frequently meet new people from diverse backgrounds. I have recently oriented a Japanese visitor about Saath's work, which was my first interaction with a foreigner. I have been anchoring the City Level Talent Show organised for the Youth Force members and the Nivea – Saath collaboration CSR event for supporting school going children."

His future goals include pursuing a course on journalism and his post graduation. He may, however, continue in the NGO sector. "I am looking for a career which would enable me to make a mark for myself rather than emphasize on the size of income and perks attached. The Youth Force platform has been a strong supporting pillar in my development and career," he said.

He started work as an Assistant Coordinator at the Research, Documentation, and Communication cell. Research Documentation and Communication (RDC) is entry point for all interns, national and international visitors, and researchers who want to associate with Saath. RDC does a multitasking job by providing research and documentation support, sending out award applications and proposals to various organizations, fundraising, CSR activities, and social/print/electronic media coordination for brand building of Saath.

Abhishek is presently working with IIM, Ahmedabad as a Programme Coordinator

Our engagement with the youth intensified from 2005 onwards. We started training young people for entry-level jobs through the Udaan and Umeed programmes[9]

---

[9] Described in Chapter 5, Page 49

In 2011, after training over 40,000 young men, we detected a disturbing trend. We found that young people from low-income neighbourhoods were developing diffident attitudes regarding their future. The idealism, curiosity, energy, anger, and risk-taking that was associated with the youth, was decreasing. Instead, we could sense insecurity and unrealistic expectations, leading to attitudes of helplessness. We found that the youth who had not been able to access higher education due to affordability or had dropped out at 10th or 12th grade and were subsequently under employed, had toned down their aspirations, which was restricting their progress.

In 2009, the Government of India set up the National Skills Development Corporation (NSDC) to push for skilling young people. Various central government ministries and state governments followed suit with a plethora of skills training programmes. Skills training providers mushroomed in the private and NGO sector. The main criteria for grants, loans, and business viability were the number of youth trained and skilled. Young people were becoming fodder for skills training at the cost of their aspirations and all-round development. Taking initiatives, involvement in sports and cultural activities, and yearning for a better quality of life was reducing amongst the youth we were engaging with. It was disquieting and, we decided to orient youth from low-income neighbourhoods towards holistic self-growth and development with the Youth Force Program.

In 2011, we consulted youth living in slums and chawls in Ahmedabad and Vadodara to understand their aspirations, worldview, and life-styles. We found that they felt that they were isolated and did not have opportunities that were required for their overall growth. This could be for mentoring, sports, recreation, and social engagement. They also expressed a strong desire for jobs that could enable a better life and, individual and social recognition. Girls felt that they were more bound by social norms than boys.

We designed and implemented the Youth Force Programme (YFP) in the cities of Ahmedabad, Vadodara, Surat, and Rajkot; and in Mumbai, in coordination with YUVA, a Mumbai based NGO. YFP created spaces and platforms for young people through which they could explore, understand and express their aspirations. They acquired skills that enabled individual growth and effective citizenship. The pedagogy was mainly experiential learning through individual and group activities, to enable discovery of innate aptitudes and talents, and developing new skills. Individual and group mentoring and counseling through experts and peers, was offered and encouraged. They were safe spaces where the youth could share their dreams, concerns, and fears. They were platforms where the youth planned numerous activities and reflected on the outcomes. Friendships and networks were created, which still continue.

A typical Youth Force Group (YFG) consisted of twenty to thirty girls and boys aged between eighteen to twenty-four years. A Saath trained facilitator was available as a mentor to YFGs. YFGs planned activities based on member's interests and needs and implement them. Each YFG would choose leaders for varying activities. Leaders who demonstrated interest in community building activities were trained to become Social Animators. Youth Activity and Resource Centres were opened in each city with a library, meeting space, computer and Internet facilities. These became safe, physical spaces where the youth could meet, discuss, plan, and reflect.

YFGs planned and participated in activities that facilitated self-growth, team building, and community engagement. These activities enabled the youth to get a wider perspective of the social, economic, and physical milieus that they lived in. With a broader understanding, they could figure out how to constructively engage with their surroundings to attain their aspirations and better their lives. By planning, organising, and managing these activities, not only did they learn new skills, but gained in self-confidence and self-esteem.

In camps and outings, the youth interacted with nature, understood effective leadership, and the benefits of mutual support through teamwork. They discovered that they faced similar challenges, which led them to share their journeys and circumstances, and subsequently, they ideated and discussed novel solutions. Cynicism gave way to hope and enthusiasm.

With the Municipal Corporation, government departments, hospitals, communities, NGOs, and academic institutions, the YFGs designed and organised campaigns to increase awareness about local issues and appropriate resolution mechanisms. They organised medical treatment and blood donation events, cleanliness drives, greening and tree plantation initiatives, voter ID and registration, awareness about HIV-AIDS, cancer, tobacco consumption and swine flu, women empowerment and girl education.

By volunteering in these campaigns, they learnt how to partner with stakeholders. They realized that they could catalyse local effective action and bring about tangible change. From passive observers, they became agents of change.

The YFGs organised seminars with resource persons in schools and colleges on topics ranging from life skills, goal setting, team-work, time management, positive attitudes and productive work ethics. With neighbourhoods, the YFGs organised community meetings to enhance awareness about education, health, entitlements, and rights.

The theme of understanding and appreciation of gender issues, cultural and diverse identities was interwoven cutting across all activities.

YFG members were encouraged to explore livelihood options which best suited their aspirations and skills. Through their association with varied YFG activities, they were able to understand the requirements for the livelihoods they aspired for. They understood the pros and cons of employment and entrepreneurship. They were able to make informed choices and, could comprehend the effort that would be needed to chart a path to earn a decent livelihood.

Urban Youth Animators

Youth who demonstrated leadership abilities and expressed their desire to become change makers within their communities, participated in the Urban Youth Animator (UYA) programme. In the UYA training, they learnt and practiced leadership styles, teamwork participation and collaboration, effective communication, problem solving and conflict management, valuing gender differences and cultural diversity, building effective relationships, taking initiatives and becoming role models, and engaging with and empowering communities and constructive citizenship.

Through these activities, YFG members were able to discover and practice their innate skills, new skills while carrying out tasks, to work as a team, understand and resolve needs of communities and engage constructively with government departments, academic institutions, employers, financial institutions, political groups, and civil society.

During the course of YFG activities, the specific skills learnt and practiced by the youth were planning, implementing and evaluation; leadership; effective communication; team work; use of technology, marketing, financial planning and budgeting.

Through peer learning and sharing, members were able to chart a course to realize their aspirations. They learnt about vision building, constructive value frames, calculated risk taking, innovation, being solution oriented, broadening their spectrum of activities, interests, being career oriented as well as nurturing hobbies. They acquired the awareness and skills to become stronger individuals, leading a better life, becoming worthy citizens and being able to give back to others less fortunate.

Through the Youth Force, twenty-four groups were formed and about 2,800 youth participated in various activities.

It is often not recognised that low-income neighbourhoods constitute a vast market. The following chapter we describe the huge potential of this market.

Chapter 11

Market Engagement

India is urbanising at a rapid pace. According to the Census of India, the urban population of India was 78.94 million (17.97 percent of the total population) in 1961 and increased to 377 million in 2011 (31.16 percent of the total population) and is expected to reach 590 million by 2030 (40 percent of total population) according to a study titled "India's Urban Awakening" by McKinsey Global Institute (MGI).

Urbanization has the potential to be one of the biggest drivers of economic development in India. The MGI's report of 2010, says that "If India pursues a new operating model for its cities, it could add as much as 1 to 1.5 percent to annual GDP growth, bringing the economy near to the double-digit growth to which the government aspires." It goes on to say that "India has a young and rapidly growing population—a potential demographic dividend. But India needs thriving cities if that dividend is to pay out. New MGI research estimates that cities could generate 70 percent of net new jobs created up to 2030, produce around 70 percent of Indian GDP, and drive a near fourfold increase in per capita incomes across the nation."

The same report goes on to say that, "In a global context, the scale of India's urbanization will be immense. India will have 68 cities with a population of more than 1 million, 13 cities with more than 4 million population and 6 megacities with a population of 10 million or more." The rapid urbanisation will pose immense challenges of water and sanitation services, transport, health, education, and housing and, safety and security of its citizens, especially women and children.

Goal 11 of the Sustainable Development Goals (SDGs) is to make cities inclusive, safe, resilient, and sustainable. The goal states that more than half of the world's population now live in urban areas. By 2050, that figure will have risen to 6.5 billion people – two-thirds of humanity. Sustainable development cannot be achieved without significantly transforming the way we build and manage our urban spaces.

Extreme poverty is often concentrated in urban spaces, and national and city governments struggle to accommodate the rising population in these areas. Making cities safe and sustainable means ensuring access to safe and affordable housing and upgrading slum settlements. It also involves investment in public transport, creating green public spaces, and improving urban planning and management in a way that is both participatory and inclusive.

Sustainable city life is one of 17 Global Goals that make up the 2030 Agenda for Sustainable Development. An integrated approach is crucial for progress across the multiple goals.

Smart Cities and JNURM

In June 2015, the "100 Smart Cities Mission" was launched by Prime Minister Narendra Modi. A total of ₹98,000 crore (US$15 billion) was approved by the Indian Cabinet for the development of 100 smart cities and the rejuvenation of 500 others. Smart Cities Mission is an urban renewal and retrofitting program by the Government of India with the mission to develop 100 cities across the country making them citizen friendly and sustainable.

The previous government had launched the Jawaharlal Nehru National Urban Renewal Mission (JNNURM) as a massive city-modernisation scheme. An estimated total investment of over $20 billion over seven years was earmarked. It was then extended by two years. JNURM aimed at creating 'economically productive, efficient, equitable, and responsive Cities' by a strategy of upgrading the social and economic infrastructure, provision of Basic Services to Urban Poor (BSUP) and wide-ranging urban sector reforms, to strengthen municipal governance in accordance with the 74th Constitutional Amendment Act, 1992.

Our experience with these national programmes have been mixed. Our main criticism is that although both programmes state, the development of the city plans have to be consultative, with participation of all stakeholders. The reality was that slum residents and lower middle-class citizens, who make up more than 65 percent of the population, were rarely consulted. The aspirations and needs of these segments were taken for granted by the policymakers and consultants, who prepared city level plans. To take an uncharitable point of view, we contend that the planners, who have got us into the mess that our cities are in today, are the ones who are developing solutions. For example, if our planners had created affordable housing options for people who were predicted to migrate to cities due to planned industrialisation, then our cities would have had fewer slums. Because of a lack of foresight, industrialisation happened, but the housing for migrants did not.

Why can't our policy makers and planners be humbler and make an effort to understand the aspirations and difficulties that a majority of the citizen's face? Who should decide whether investments should be made in public transport or on roads used mostly by private vehicles? Who should decide that it is acceptable to disrupt livelihood and social networks that are created by slum residents, over many years, by relocating slums? Who should decide to reduce investment in public health and education in favour of private service providers? Who should decide what a smart city should be?

It is our fear that India's drive to create smart cities will not be inclusive. The bias towards technology, as a driver for efficient management of our cities, will alienate the urban poor who are not technology-savvy. Will smart algorithms decide the public transport routes or will the needs of the less privileged, children, and the elderly be considered?

We contend that mainstreaming people living in low-income neighbourhoods provide a solution to making our cities livable and vibrant. Because of their large numbers they represent a huge market. They have high levels of aspiration and generate adequate surpluses to spend on education, health, housing, as well as other goods and services.

Slums, chawls, and low-income neighbourhoods offer a huge opportunity for both, for-profit and not-for-profit social enterprises.

Opportunities in low income neighbourhoods

The size of the market representing low-income neighbourhoods is quite attractive. The census of 2011 recorded 1.3 crore slum households. This population was estimated to surge to 10.4 crores by 2017 – or around 9 percent of the total projected national population of 1.28 billion that year. It means 2.1 crore households with 5 members per household. If compared to the population of countries, the slum population in India would rank as the 14th largest country in the world.

Is this a problem or is it an opportunity for social innovation, economic growth, and sustainable business? What do these numbers mean?

At an average, HH spend of Rs 10,000 per month, it is an annual market of Rs 2,52,000 crores; Rs 2,520 billion or Rs 2.52 trillion.

Here is a further breakdown of the household expenditure in services:

Service	Spend per month in Rs	Spend per month of 2.1 crore households in Rs Cr	Annual spend of 2.1 crore households in Rs Cr	Annual spend of 2.1 crore households in Rs Bill
Electricity	200	420	5,040	50.40
Education	1,000	2,100	25,200	252.00
Health	1,000	2,100	25,200	252.00
Rental	2,000	4,200	50,400	504.00
Financial Inclusion	1,800	3,780	45,360	453.60
Food	3,000	6,300	75,600	756.00
Others – white goods, vehicles, etc	1,000	2,100	25,200	252.00
Total	10,000	21,000	2,52,000	2,520.00

Is this not an attractive business proposition? Yes it is. Can it lead to sustained economic growth? Yes, it can. But social innovation is required to crack this market. It cannot be cracked through traditional business methods.

At Saath, we understood the potential of this market, when almost 5,000 slum households in Ahmedabad spent Rs. 1 Crore to get basic services in the slums they lived in and then spent Rs. 1,250 Crore to upgrade their houses. 2 lakh households spent Rs. 40 Crore to get electricity connections. Almost 5,000 affordable houses were bought at a total cost of 500 Crore. Our Saath Credit Cooperative Society gave loans amounting to Rs 16 lakhs for housing upgradation in slums.

We saw that slum resident's preferred sending their children to private schools and paying fees, rather than send them to inexpensive government schools. Similarly, they chose to pay fees to private health facilities rather than going to government health facilities.

When articulated the potential of slums being a market to the government, business community, and the civil society, we were not taken seriously. The government's decades-old welfare driven attitude was difficult to change. The private sector catered to a huge middle-class market and were wary of taking risks. The NGO community was divided over ideology and, the academic community asked for data.

We went to our drawing boards to design a model that would demonstrate the potential of the market. We also coined a new term – the Urban-Not-So-Poor (UNSP) to characterise the market segment that had high aspirations and purchasing power. Classically, this would be a segment between the poor and the lower middle-class.

The model that we designed was the Urban Resource Centres.

Urban Resource Centres (URCs)

The URC was designed as a single multipurpose window for demand and supply data aggregation of services and goods required by the UNSP market, and then shared the demand data with various service providers, including government departments, private sector companies, social enterprises and NGOs. Later, academic institutions found the URCs to be a convenient place to gather data for academic and business application.

A typical URC is located in a slum cluster of about 5,000 households. The staff of the URC consisting of a coordinator, data manager, and three field officers, gather data from households. The data included demographic profiles, household status, occupations, financial inclusion, disability and special needs, documentation levels and awareness of government schemes and entitlements. The data is then segregated and shared with various service providers.

Some examples are:

a. Identifying the households that did not have a toilet and sharing it with the concerned government department, which would then follow up by building a toilet
b. Providing skills training providers with data on youth who wanted jobs. The skills providers contact the youth, train them and get them jobs.
c. Identifying people with disabilities and connecting them with NGOs and government departments for certification, as well as helping them get wheelchairs and other equipment
d. Identifying people with inadequate documentation, like PAN Cards, Aadhar Cards, Caste certificates, BPL certificates, ration cards and then facilitating procurement of such documentation.
e. Identifying the financial inclusion requirements and then connecting the households with banks, MFIs and insurance companies

These are a just a few examples of the potential of the URCs.

The more significant role of the URCs is that of defining the characteristics of the market segment that low-income neighbourhoods represent. The data on expenses show what the expenditure priorities of this segment are and, the average amounts spent. Conversely, the data shows the purchasing power of this segment and the surpluses generated. This data can be very useful for the private sector, government, and civil society

Through the Urban Resource Centres, we have reached out to almost 56,000 households and facilitated linkages for about 38,262 persons.

Disasters, both natural and man-made impact and cause significant distress in communities. The next chapter describes how Saath has supported communities affected by disasters and displacements.

Chapter 12

When Disasters Strike

The Gujarat earthquake

On 26th January 2001, Gujarat was devastated by a major earthquake measuring 7.7 on the Richter scale. The epicenter was in Bhuj, the capital of Kutch. Ahmedabad, Rajkot, Jamnagar, Surendranagar and Patan districts were severely affected. **The earthquake killed almost 20,000 people, injured another 167,000, destroyed nearly 400,000 homes and about a million people were rendered homeless.**

Our immediate concern was the well-being of communities we worked with in Ahmedabad. For once, the ramshackle houses in slums proved to be a blessing in disguise. Most houses, not built with concrete and single storyed, survived. The ones that had collapsed did not cause serious injuries or deaths. There had been considerable damage to other structures, as well as deaths, in Ahmedabad. The government, civil society, and individuals in Ahmedabad responded immediately with relief materials and services. With the overwhelming response in Ahmedabad, we wanted to find out the situation in Kutch, where the damage was devastating and, the support systems by the government and civil society weaker.

Kutchchh, because of its location, has a long history of disasters. In 1998, a super-cyclone had hit Kutchchh and caused considerable damage. Kutch Nav Nirman Abhiyan (KNNA), a network of civil society organisations, had been formed to assist the government and aid agencies in coordinating relief and rehabilitation for the victims of the cyclone.

On 29 January 200, we visited Bhuj to assess the damage and explore our role in providing relief. We met representatives of KNNA, who were coordinating the earthquake relief efforts. They invited us to join them, as there was a dire need for organisations to coordinate relief in the affected rural areas of Kutchchh. We mobilised a team of twelve volunteers from Saath who arrived in Bhuj on 5 February 2011 to undertake relief operations.

We were given the task of carrying out rapid assessment of damage caused in the villages of Bachau taluka, one of the most affected areas, by KNNA. This meant surveying all the 70 villages in the taluka to determine the number of people killed, or injured, the number of houses damaged, the number of grazing animals killed or injured, the extent of damage to the water tanks and, the type and amount of emergency assistance received and further required.

Emergency Relief

During the two days of rapid assessment, we witnessed, first-hand, the devastation caused by the earthquake. This led to our decision to be involved in at least the relief stage, if not longer. We observed that villages that were further away from the main roads and administrative towns received much less assistance. This was especially so in the villages of the Khadir area bordering Pakistan. We submitted the rapid assessment report to KNNA and informed them about our decision to join in, in the relief operations. KNNA gave us the option of managing one of the Emergency Response Centres being set up across Kutchchh for coordination of relief work of the government and aid agencies. We selected the Khadir area for setting up of the centre. This was the beginning of a four-year relationship, not only in the Khadir region, but also in Rapar town.

Khadir is a unique region in Kutchchh, which itself is a unique district in Gujarat. The region stands out differently in terms of its location, physiographic characteristics, social ecology, economic systems, and culture. The region is isolated, it is an island in the Greater Rann of Kutch and is linked to mainland Kutchchh through a single bridge over a creek at Amrapar in Khadir and Lodrani in the Kutchchh mainland. The region, because of its isolation has remained at subsistence economic level. Ecologically, Khadir forms part of the Great Rann ecosystem. It is better known as the region where the Harrappan ruins of Dholavira are located.

The Khadir area comprises twelve villages and five hamlets, with a population of 11,000, 2,200 households, and a diverse social structure. Kolis, the largest community (28 percent) are landless and work as agricultural labourers, saltpan workers, and as construction workers in other parts of Kutchchh. Ahirs (23 percent) the landholding community keep cattle and have started diversifying into new business activities. Darbars (11 percent) are also landholders; women from this community do intricate embroidery work. Harijans (8 percent) are landless and work as agricultural labourers. Rabaris (5 percent) are cattle owners. Muslims (5 percent) work in a variety of activities, from farming to transport. Kolis and Ahirs are seasonal migrants to other parts of Kutchchh during the summer months.

We set up an Emergency Response Centre (ERC) at Ratanpar village on 9th February 2001, which were basically a few tents and a team of six people. From the ERC, we distributed relief materials provided by the government, relief agencies, and individuals. We undertook surveys to assess the damage to houses and water storage systems, and the situation of fodder and water supply. We also checked the functioning of the Public Distribution System (PDS); whether cash subsidies provided by the government went to the target beneficiaries.

We conducted a detailed survey of the old and destitute, widows, handicapped, and injured persons, as well as of those who lost their lives. This survey formed the groundwork for the government's relief and aid packages. We conducted a household survey at the village level for a better understanding of the socio-economic status of the residents. The approach of the ERC team was to work in partnership and consultation with existing village governance systems. This involved calling of gram sabhas (village meetings) for all major decisions, to ensure transparency and accountability. We recruited five youth from the villages into the ERC team.

February is very cold in Kutch; we were able to procure and supply blankets and warm clothing through an aid agency. Availability of drinking water was restricted in most of the villages, as the water structures had been damaged. We approached the Gujarat Regional

Water Supply and Sanitation Project to make an assessment. They recommended building of water storage systems with separate storage and supply areas. The size and number of storage systems would depend on the demographic profile of the village. We discussed the recommendations with the village committees. It resulted in a partnership in which we would contribute water tanks, plumbing materials, cement, and cost of masons and plumbers. The villagers would contribute stones and sand for the construction, and labour assistance to the masons. The village committees would decide location and size of tanks, where the main selection criteria was proximity and availability to all village residents. We constructed 30 water supply systems in various villages with assistance from Cordaid and, set the tone for a longer partnership with the affected people and aid agencies. Similarly, we worked in consultation with marginal and small farmers to provide seeds to almost 1,500 farmers for the monsoon crop sowing.

Our assessment for shelter indicated that almost 200 houses would have to be rebuilt. The remaining were in need of renovation and repair. However, the impact of the earthquake and the subsequent fear and anxiety, led to a situation where people were afraid to live in their houses although they were declared safe.

With summer approaching, which is very hot in Kutch, KNNA decided to provide interim shelters to almost all the affected households. The assistance was in the form of a construction kit, which comprised of tiles, poles, and cement. Local stones were to be used for constructing walls of about 3 feet high, which could not cause damage even if they fell. Village residents were trained as masons for constructing the basic structure. We ensured distribution of the interim shelter kits in Khadir, as well as training of masons. We supplied 1,200 interim housing shelter kits in Khadir

By June 2001, we had a better insight of the vulnerability of people living in Khadir from a social, economic, and environmental perspective, based on our socio-economic survey, continuous physical presence, and interaction with residents.

These were our conclusions:

a. The Scheduled Caste and Scheduled Tribe communities in the area are amongst the most vulnerable groups in Kutchchh District, with limited access to basic services and facilities.
b. Due its location and extreme climatic conditions, the communities have inadequate access to opportunities for improving livelihood, enhancing literacy levels, and improving their health status.
c. The common property resources, in terms of grazing areas and water resources, are rapidly dwindling.
d. The communities have economic links with other parts of India through their handicraft work, for which there is substantial demand. However, they do not get commensurate returns from the middlemen who facilitate the purchase and marketing of their goods.
e. The women are deeply rooted in their social norms and behaviour patterns. One such devastating instance – a group of women did not leave their building during the earthquake and ended up buried under it.

On the basis of these findings, we planned our rehabilitation strategy. The main aspect of the strategy was that we would work with them for a period of at least three years. The integrated development approach would be participatory. The activities would include:

i. Construction of earthquake resistant houses, water storage, sanitation failicities, and rainwater harvesting structures
ii. Afforestation and environmental conservation
iii. Promotion of self-help groups for savings and credit
iv. Training in various skills and support to entrepreneurs
v. Increase health awareness
vi. Support pre-school classes through training of teachers with appropriate teaching aids
vii. Increase disaster preparedness
viii. Build capacities of local groups to sustain the activities

Cordiad, a Dutch aid agency, with whom we have had a funding relationship since 1993, agreed to support the integrated rehabilitation of Khadir.

Housing and Shelter

Building earthquake resistant houses was a priority for residents whose homes had been destroyed by the earthquake. We decided to support construction of houses for vulnerable social communities and individuals – the widows and handicapped. We held intensive consultations with the residents about types and designs of houses. The design had to be suitable for the semi-arid environment of Khadir, easy to build and can be repaired locally. It also had to be earthquake-resistant and in keeping with their social norms.

We consulted architects and engineers to discuss designs and structures. Finally, we zeroed in on housing design options. The first was a traditional design using stones, cement, and roof tiles. The second one was a modular construction of reinforced cement. Both designs incorporated living spaces, kitchen, courtyard and, a bathroom and toilet. We built sample houses of both designs at our campsite and invited residents to inspect and choose their preference. The residents were canny and tested both the designs to the extent of ramming a tractor into both the houses to ascertain the strength and durability. Finally, they opted for the modular reinforced concrete design. The traditional design was equally safe, but a sub-conscious fear prevailed, as many similar houses had been destroyed during the earthquake.

We got the modular design approved by the government authorities and started production. The cost of construction was met partly through the subsidy given by the government and the balance, through a Cordiad grant.

The LEGO type design of the modular houses required a production centre where the various components would be manufactured, as per design, and the components then transported and assembled on site. With the help of Mr. Kiran Gandhi, the architect whose design we had adopted, we established a production centre at Ratanpar and commenced manufacturing. The local people remarked that this was the first time a manufacturing unit had come up at Khadir. They came and saw the production process. It increased their confidence in the structural strength of the houses. We contracted the assembling of the houses to specialised contractors from Ahmedabad. They, in turn, trained and employed local residents and house-owners as masons for the work.

Taking into account the social dynamics of Khadir, we had to consider the various social and traditional relationships. This meant that the employment generated out of the production and assembling of the houses had to be shared equally between the different social groups. It was difficult to maintain a balance, considering the need for quality work, timely construction, as well as the budgetary constraints. In this mammoth and complex task, we built 176 houses in Khadir between March 2002 and June 2004.

Water resource management and afforestation

Khadir is a drought prone region with scanty annual rainfall of approximately 48 cm, which often fluctuates between 14 to 74 percent of the normal. The frequency of a drought is once in every 2.5 years. Our approach was to create water security by upgrading existing check dams or building new ones, desilting existing farm ponds and creating new ones, digging wells and recharging existing ones, as well as supporting the construction of contour bunds to reduce erosion of good topsoil. Two new check dams were built and nine repaired. Two hundred and fifty wells were dug to facilitate recharging of the underground water sources. Four large village ponds were desilted, and the silt was used in the farms. In almost all the villages and hamlets, we supported the construction of contour bunds, which helped retain rainwater. Drip irrigation systems were installed in three villages.

The criteria for selecting villages and farmers were their readiness to contribute, either in cash or kind, towards the cost of construction. If the construction benefitted a larger public, the contribution was 10 percent of the cost. If it benefitted individuals or a group of farmers, the contribution was 25 percent. Most of the contributions were in the form of labour or supply of stones, which were an integral part of construction. We facilitated the provision of cement, sand, and mechanised construction equipment, along with technical experts. Village and farmer committees were formed, which approved the civil engineering work and estimates, monitored the progress, and ensured payment of bills for the work done.

We worked with the Gujarat government's forest department for free saplings of various trees that could grow in the harsh environment. Trees were planted in all the school compounds and in other public spaces. Saath bore the transportation costs; the community nurtured the saplings.

Education and community groups

Sixteen day-care centres were established in all the villages. We built twenty-four centers and the remaining were built by other agencies. The day care centres followed the Balghar methodology, in which, women from the respective villages were trained as teachers. We provided nutrition support and teaching materials. Subsidised fees were charged. Due to strong caste differences, we sometimes had to have separate day-care centres for the Harijan community.

Mahila Mandals in all the villages monitored and supported the day-care centres. Some of the Mahila Mandals became self-help groups, providing savings and credit services.

Youth groups were formed to facilitate and monitor ongoing work with regard to water resource management and construction work. These groups also started sports activities.

Rapar Town

Rapar town is a Taluka headquarter with a population of 28,000, according to the 2011 census. It is the gateway to the Khadir region. Immediately after the earthquake, we were actively involved in the distribution of relief material in Rapar town. The poorer families in the town bore the major brunt of the devastation, lives, properties, and livelihoods collapsed.

On completion of the relief phase, we set up base at Rapar to work out strategies to ensure the urban poor were included in the affordable shelter and basic infrastructure facilities in the new town planning scheme. The Rapar Area Development Authority (RADA) was established to implement the rehabilitation package announced by the Gujarat Government. Financial assistance by the government for construction of a house was pegged at a maximum of Rs. 55,000. The urban poor would be given plots measuring 25 sqm. at an earmarked location (within the town plan zone), with an option of purchasing an additional 25 sqm at a highly subsidized cost of Rs. 2,500. Those who owned land would be given Rs. 55,000 for in situ upgradation. They would be legal owners of the houses.

RADA allocated a large piece of land to accommodate families to be relocated. Based on its interaction with the affected below poverty line families, we decided to build 200 earthquake resistant houses at the relocation site and, 50 houses at in situ locations. We approached Mr. Rajendra Desai and Ms. Rupal Desai from the National Centre for People's Action In Disaster Preparedness (NCPDP), an Ahmedabad based NGO, which works towards bringing viable, eco-friendly and sustainable technologies to help people reduce their vulnerability in the aftermath of disasters, to prepare a comprehensive plan in all matters related to design, architecture, civil and structural engineering, and construction for the housing rehabilitation

To increase the accountability of all concerned, a memorandum of understanding (MoU) between the beneficiary families, Saath, RADA, and the Deputy Collector's (Earthquake Relief) office was drawn up, which listed the project commitments that the concerned government departments and the NGO had to honor. The beneficiary family made a commitment of a financial contribution to the project, which was almost 70 percent of the earthquake related compensation. The contribution gave them a sense of ownership once the shelters were built and occupied. The money contributed in excess of the cost of the house was used to build more shelters and cross subsidise families who received negligible or no compensation at all. We would like to note our appreciation for the RADA authorities who were very cooperative in this effort the provide housing for the very poor.

Unlike Khadir, these families opted for a traditional design of brick, tiles, cement, wooden beams, and rafters. The houses had two rooms and a sanitation block. Beneficiaries trained as masons were involved in the construction of their houses. The RADA authorities that subsequently released the housing subsidy approved the constructed houses. The resident's need for a community hall was fulfilled.

Two years after the construction of houses in Rapar, we did an evaluation to assess the condition of the houses. There was one anomaly – the wooden centre beam had bent. The NCPDP team devised a mechanism that successfully addressed the problem.

Rehabilitation in Maliya Taluka

Maliya taluka is a part of Morbi district in western Gujarat surrounded by the Arabian Sea, Kutch, Rajkot, and Surendranagar district. Morbi district was severely affected by the earthquake.

We were not involved in the immediate relief phase after the earthquake. We participated in Care India's multi-dimensional earthquake rehabilitation programme – SNEHAL (Sustained Nutrition, Education, Health and Livelihood Project), in Kutch, Rajkot, Patan, and Surendranagar districts of Gujarat. The SNEHAL programme aimed at achieving livelihood security of the poor and marginalized, mainly primary producers of salt, and charcoal, fishermen, and milk suppliers, as well as weavers and handicraft workers. We worked in Maliya taluka and Surendranagar district.

Maliya Taluka

With Care India, we conducted a participatory assessment (PRA) in 60 villages in Maliya taluka to identify villages most vulnerable in terms of livelihood security. Saath subsequently worked in the 15 most vulnerable villages. We established an office in Morbi town and recruited a team of 8 persons, which included people in the fields of engineering, health care, animal husbandry, as well as livelihood specialists.

We initiated our work in all the villages by forming grass root committees of farmers and women. These committees were partners in decision-making, implementation and monitoring of all village level interventions.

We discussed the PRA assessments with these groups to help decide which works related to water, farming, animal care, land levelling, check dams, bund contours, drip irrigation, etc. should be taken up in each village. The committees, after consultation with individual producers, decided the works to be done through a formal resolution, which was handed over to us. These works were partly funded by CARE. The work would then be implemented through the committee with specialist support from us.

In Kakrechi village, the village committee had recommended construction of a check dam within a budget of Rs. 12 lakhs. The villagers contributed Rs. 5 lakhs and CARE and the Coastal Salinity Prevention Cell contributed the balance amount. Our engineers supervised the construction. Material was purchased through tenders by the committee.

With a history of having been washed away six times, the dam constructed by the committee and Saath still stands.

Deepening of existing village ponds or excavation of new ones was taken up in a similar fashion, with financial contributions shared between the villagers and us. Contribution by villagers was mostly in kind – by way of labour or use of their equipment. The farmer committees approved all bills for final payments.

Small farmers without access to irrigation facilities or bore wells excavated farm ponds and lined them with plastic to prevent seepage and sprayed them with a monomolecular film to reduce evaporation. These farmers were able to grow vegetables as an additional cash crop.

We worked with the farmer committees to facilitate regular farm animal health camps and immunisations in partnership with veterinary colleges. These were especially effective in controlling foot and mouth disease during the monsoon.

With the women's groups, health camps were organised for fisher folk and the Agariya community who suffer severe skin diseases; a result of working as salt pan workers in very harsh conditions.

The women's committees ensured quality of the mid-day meals served in schools and Aanganwadis. They helped improve school facilities by influencing teacher committees and village Panchayats. They addressed the pressing issue of child labour (when school children dropped out of school during the cotton harvesting season) by creating awareness among the parents.

In Haripar village, women formed Self Help Groups (SHGs). These groups were provided with a revolving fund of Rs. 60,000 to reduce spurious borrowing of working capital from money lenders who lent them on the condition that the salt was sold to them at almost 25 percent of the market price. The SHG lent its members at 5 percent annual interest and, provided mutual guarantees for repayment. Similarly, fishermen groups were provided with a revolving loan as working capital to buy and repair nets, for storage facilities, and purchasing equipment.

The women's groups from the fifteen villages formed a federation for taluka and district level advocacy.

An Agriculture Resource Centre (ARC) was set up in Kakrechi village, for which the Panchayat provided land and the farmer committees contributed construction material. The ARC consolidated the input needs of farmers to enable them obtain supplies at affordable prices. They also guided farmers on appropriate use of fertilizers, seeds, and pesticides.

A Rural Resource Centre was established in Maliya town to facilitate access to the numerous government schemes for vulnerable communities.

Surendranagar

Surendranagar, another earthquake affected district, is home to a substantial number of artisans and weavers engaged in khadi, patola and tangaliya weaving, as well as brass handicrafts. Tangaliya is a beautiful, but tedious art of weaving (used to be only in wool), which dates back 700 years and is particular to the villages of Surendranagar District in

Gujarat. There used to be a large community of Tangaliya weavers but now the numbers have dwindled.

Babubhai Rathod is from Dedadra (in Wadhwan Taluka of Surendranagar District) and is an expert Tangaliya weaver. He told us about his craft and how they used to weave only in wool that was hand-spun locally. Their customers were Bharwad women (shepherd tribe) who only wore these woolen long skirts woven with traditional motifs. But as time went by, and in Babubhai's words, "As the "fashion" era came in, and colourful, mill woven, cotton material came into the market, the Bharwards moved with the times; changed over to wearing cotton instead of woolen." The transition took ten years, during which time their incomes started dwindling and weavers gave up their traditional weaving and went into other trades.

It was during these trying times that Saath got in touch with them. They didn't know anything about who or what Saath was; neither did Saath know anything about them. They had heard about the Tangaliya art of weaving and had come in search of them. They, along with students from the National Institute of Fashion Technology (NIFT) and representatives of CARE worked out a training programme for them. They were taught how to use cotton in their art. They also introduced them to different colours and designs.

They were given stipends during the training period. The youth were organised into a working group and were made aware of how to go about saving. They went through periods of ups and downs for a while.

The village went through a water crisis as the village water tank gave way. Saath helped build a tank and when it gave way some years later, they once again helped rebuild it.

Babubhai has been associated with Saath for the past eight years. He goes on to say that working with Saath has been good. He sells his products to Rweaves at Saath and gets a good return. He is very grateful to them because they have guided him in terms of colours, patterns, and products.

When he was the lone weaver supplying goods, he found it difficult to produce in good quantity due to paucity of funds. Saath put him in touch with an organisation, Rang De, which assists artisans with loans. He recalls his first loan of Rs. 40,000 and then of Rs. 2,00,000. It helped set him up for full-scale production. He has not had to look back since; it has been a forward move to becoming self-sustaining. He attributes it to Saath's way of encouraging people to work towards independence.

He went on to say that he has never had a bitter experience of any kind with Saath. He also adds, that Saath invites them to participate in other programmes and exhibitions as well as live demonstrations. He started out working alone to meet production orders. Today he has a group of six to seven weavers affiliated with him. They too had started from scratch and are doing well for themselves.

On being asked what specifically did Saath do to bring about this turn-around, he explained that Tangaliya originally was not the fabric or the pattern, it was the garment itself; the skirt that the Bharwad women wore. But now Tangaliya represents the dots that make up the design or pattern. He further went on to explain that, the Dangasiyas, the community he belongs to, were originally Bharwards. His story goes – "Bharwads lived in community settlements with space around each house for their cattle. One day, a calf died and the younger of two brothers picked up the carcass and disposed of it. On his return, the elder brother told him not to enter or to touch anyone as he had handled a dead animal. He moved out and lived separately. This branch of the Bharwads is today's Dangasiyas." According to Babubhai, the name Dangasiya comes from the word Dang, the wooden rod that the Bharwards carry to herd their cattle.

The art of weaving Tangaliya is a craft that all Dangasiyas are comfortable with. The original ethnic designs and patterns are still used in some products, along with the modern designs incorporated after their training.

Saath has helped bring Tangaliya, an unknown, disappearing art, onto the international map. People go to visit them of their own accord now. Babubhai said, "Our income levels have gone up 40 to 50 percent and our art is now flourishing. I am very thankful to Saath for all of this."

With the Snehal programme of CARE India, we helped form 155 Self Help Groups (SHGs) of artisans in twenty-five villages. We then facilitated the formation of the SUVASA Federation – a federation of all the SHGs.

The National Institute of Fashion Design (NIFT) joined as a partner. It provided professional inputs on brand building of SUVASA, introduced contemporary designs and products, and created marketing links with larger wholesale and retail enterprises. A Geographical Indication (GI) tag, which certified the uniqueness of Tangaliya weaving in Surendranagar district, was obtained.

With SUVASA, a raw material bank for purchase of cotton, silk, and wool was established to ensure the weavers could buy raw materials on credit at affordable rates. Through SUVASA, Dena Bank and State Bank of Saurashtra graded artisans for financial assistance. Artisans were linked with government welfare and insurance schemes. A revolving fund was set up to provide credit for purchase of weaving equipment. A Common Facility Centre was established in Surendranagar town to facilitate the activities of SUVASA, Saath, and NIFT

What we learnt about relief and rehabilitation after the Gujarat Earthquake

We worked in two distinct geographies during the relief and rehabilitation of the Gujarat earthquake. One was the Khadir region in Kutchchh, and the other was mainly in the Maliya taluka and Surendranagar district of Saurashtra. Our experiences were dissimilar in the two locations.

Khadir is an island surrounded by the Rann of Kutch. The inward-looking mindset associated with isolated communities was prevalent in Khadir. The residents were wary of outsiders and did not repose trust easily. As it is a backward region, there had been very few development initiatives by government and civil society, which generally lead to the formation of community and social groups. The existing groups were rigidly divided into two main caste-oriented formations. The Ahirs, whose main occupation was farming and trading, led one. The Darbars, who are the traditional ruling and warrior community, led the other. The Harijans and the

Muslims would align themselves with one of these groups. The division had also manifested itself by these groups' identification with different political parties. The caste and political divisions affected almost all aspects of economic activities in Khadir.

The limited investments made by the government led to a scarcity mindset, and as a result, intense competition between the two groups for garnering the resources we brought into the region for relief and rehabilitation. We could not always make decisions based on the vulnerabilities of a community as it led to conflicts and in one instance, violence against one of our workers. We had to balance our assistance and investments between the two groups. For example, when we required tractors and other machinery for building dams, we had to be very careful that we hired these from both the groups. We could do very little work with women because the gender rigidities, due to lack of exposure to developmental activities, were very strong, compared to similar rural areas in Gujarat.

In contrast, working in Surendranagar district was less stressful. Surendranagar is not isolated as two major highways and a railway line cut across the district. There have been subsequently, more government and civil society interventions. The people were more outward looking and aware of their rights and entitlements. There were more self-help groups we could engage with. The diversity of economic activities meant increased livelihoods and income. There was more interaction between differing castes and communities. For example, when a check dam was to be built, villagers put aside their social differences and contributed for the good of the village.

Gujarat Riots

"All that I am today is because of Saath" – Noorjahan Diwan's story

"In 2002, about a fortnight after the riots, before I joined Saath, I vividly remember visiting the Vadilal hospital for a medical problem with my sister-in-law. The doctors there referred to us a 'bombs.' I was so scared that I insisted we leave without consultation. Outside the hospital we were once again chased by a mob. We got into an auto rickshaw and asked the driver to just drive us to Pravinnagar. Once we were on our way, the driver asked us point blank whether we were Hindus or Muslims. We told him we were Muslims and needed to go to Juhapura. He asked us to get off. We pleaded with him and he said he would drop us off a little away from Juhapura. He was compassionate and dropped us off and didn't even take the fare from us.

"I was very traumatised and actually felt that the Hindus were bad people. I had been living a very protected life and was always surrounded by Muslims. I had had no interaction with Hindus.

"Some time later, I visited one of the camps and was horrified at the situation. I was so shaken that I broke down and cried. When I returned home, I described what I had witnessed at the camp to my husband. He got very angry and slapped me saying that there were men there. During that time, we lived in a very strict environment at home – we wore the 'parda' (covering of the face with a cloth).

"The next day, however, on the pretext of shopping for vegetables, I visited the camp once more and I met a friend of mine who was also a teacher. I enquired as to what she was doing, and she informed me about Saath. She was working with children through a Saath programme. I told her I too would like to work there. She made enquiries and I was called for an interview. I told the interviewer that I wanted to work with the riot victims and that I would resign from my school job. She outlined the work, which included teaching children and dispensing medicines to women.

"I joined Saath the next day. I started work with the children, doing different activities with them. The children were a very traumatised lot. They suffered nightmares and woke up screaming from their sleep. They were initially petrified of Hindu women who wore bindis (a red spot on the forehead). Gradually, as the Saath team of Hindu men and women interacted with them and helped them, they started accepting them.

"After some time, we were asked to work through Aman Samudhay, an organisation working for the riot affected. We formally joined them through Saath. Here we were trained in dealing with trauma, teaching children games, how to lodge FIRs at police stations, etc. Members of the Saath staff too attended these trainings.

"Subsequently, we started work on various fronts. My first experience at filing a FIR at the police station is a very memorable one. I had gone to the police station wearing a burkha. The police officer on duty asked me to go and remove my burkha and then come to file the FIR. On enquiring why, he said, "you may be hiding a bomb underneath." I complied and then filed the case.

"I had problems on my home front too. I had two small school-going children and one six-month old infant. I was constantly reminded that my place was in my home not out in the streets.

"When the camps were being disbanded, we worked for the rehabilitation of the people. We helped distribute home kits, as well as livelihood related goods and equipment. We attended Saath meetings and learnt about peace and justice. We worked alongside them.

"People wanted to return to their homes in the Naroda Patia area but were afraid of what was in store for them. A group of us were posted in the area to assist them. We were chased by the people of the area a couple of times and had to run for our lives.

"A Muslim committee, as well as other NGOs, also assisted in the process of rehabilitation in various areas.

"Saath later started forming Self Help Groups and were encouraging women to save. I worked with them, as well as with the children. This work I did under the aegis of Aman Samudhay. I worked with them for five years, during which time I also did some work on peace and justice, towards unity between the two communities. Our centres were situated between the two community settlements. We worked to bring about unity through children, as the grownups had strong views and had formed a divide between the two communities. We encouraged the children to intermingle during studies, as well as during activities. We ensured that team games had a mix of both communities on both sides. Gradually, the children mixed freely.

"When we held parent's meetings at the centre, both communities would form their own group and take part. We took the women on an excursion and played team games with a mix of both communities in each team. We ran an embroidery centre and encouraged women from both communities to mingle. It was working. The biggest hurdle was getting the men together; they never took part in the unity and peace meetings.

"As it happened, the Torrent Power Company had started a drive to legalise the electricity connections and went about disconnecting illegal ones. It was during this time that we took up the work of assisting those in need of connections. Everyone had to come to our centre to apply or make payments. It was then that the men folk finally started coming together. Then Saath assisted in bringing the basic amenities to the area; roads were built, and water supply made available. It was the meetings held during these works that helped completely break the ice among the men between the communities. Youth from the VHP and RSS cadres in Guptanagar joined the unity campaign.

"When the political leader, Haren Pandya, was killed Muslim houses were targeted, many burnt, and people started to flee the area. The Hindus from the area formed a human chain and convinced the Muslims that they would stand by them and let no harm come to them. It was the most heartwarming time. It brought me immense joy that our efforts at unity had borne fruit.

"This, however, as was to be expected, did not go down well with the higher ups within the Hindu organisations. They came to our community centre and demanded that it be shut down and threatened to take me away if I did not comply. I stood firm. They filed a case against me at the police station claiming that I was a terrorist inciting the people. I received summons to present myself at the police station. From the Hindu Bharward community, the men came along with me and gave their statements stating that I am working on communal unity and peace. This happened a few times and, each time they came and bailed me out.

"At my home front however, they were worried that I could be attacked on my to or from the centre, so I decided not to take the risk any further. We shut down the centre.

"Our work all over the city continued. Wherever there were riots or unrest, we, from Aman and Saath intervened. We held hands and sang songs of unity. We intervened at Gomtipur and Bapunagar. We received complaints from both communities, which we solved through compromises and talks on unity. We had about 35 centres all over the city.

"We joined hands with Jan Sagharsh Manch, run by Advocate Mukul Sinha, to fight against innocent Muslim boys being jailed under POTA. During this period, we were often incarcerated.

"I later joined the Bhartiya Muslim Mahila Andolan, working with Muslim women at the state level. I helped form groups of Muslim women all over the state – Kutchchh, Junagadh, Jamnagar, Bhavnagar, Rajkot, Baroda, Sabarkantha, Banaskantha, Panchmahals, etc. I was incorporated into the national level team through which I worked in many states and later went on to become the convener at the national level. Soon I was encouraged to join a political party. I did. I joined the Congress Party and was made the state president of the Minority Cell. I created Muslim Women's groups within the party.

"I had joined Anhad in 2003 and am still associated with them. They work on peace and justice and keeping democracy alive. I am presently working with women on these issues, as well as on our constitution. I work with and for the exploited and downtrodden. My work includes working on the rights of the Dalits. Basically, I now work wherever there is unrest and exploitation.

"In the field of education, from 2003 to 2010, I have helped enroll around 500 poor children in hostels. Anhad meets their expenses through donations received for the same. Now, when a young boy or girl comes up to me, says hello and asks if I remember them, and introduce themselves as children I had helped admit into a hostel, it gives me great joy.

"Counseling married couples on the verge of a break up is part of what I do today. Our counseling centre is in Juhapura. We had run a Shariat Adalat through the Bharatiya Muslim Andolan. We mainly got cases of triple talaq. We had taken out a rally against this tradition in Juhapura and had received a lot of flak. It was after this that the issue went to the Supreme Court through a petition signed by seven people; I was one of the signatories on behalf of Gujarat.

> "I have been the chairperson of Aman Samudhay thrice. I am a recipient of a number of awards over the years. My first award was for my work in peace and justice around 2010–11. All that I am today is because of Saath. Their initiation, trainings, guidance, and nurturing have made me who I am."

On 28 February 2002, we were celebrating Saath's thirteenth foundation day at our office. During the festivities, we received calls from Sankalchand Mukhi ni Chali and Guptanagar informing us that rioting had erupted in Ahmedabad in the aftermath of the burning of passengers in the Sabarmati Express. Our staff members were requested to talk to the community leaders to prevent violence in the areas that we were working in. Some of us tried to visit the areas, but were prevented, as curfew was declared in those areas. The reports from our community staff members were not encouraging. The scale and intensity of the violence was beyond belief. Authorities were not responding to their pleas for assistance. In the ensuing days, it was apparent that the rioting, violence, arson, and looting, especially in Muslim areas and business establishments, were on a very large scale. We learnt that the victims of the riots, mainly from the Muslim community, had fled to relief camps. We took the initiative to provide immediate relief to people living in these camps; and later joined in rehabilitation efforts.

This was not the first time we were getting involved in relief and rehabilitation of riot victims. In December 1992, after the demolition of the Babri Masjid in Ayodhya, riots on a much smaller scale had taken place in Ahmedabad. Economic activities had come to a halt, as curfew had been imposed to prevent further violence. The hardest hit were daily wage earners, who often do not have savings to buy basic foodstuffs. With the support of Oxfam, we distributed food grains to almost 2,000 households in Sankalchand Mukhi ni Chali and Guptanagar in 1992. The modus operandi was working with community leaders to identify households that were the most affected. Subsequently, coupons were distributed, and we managed two distribution centres with the aid of the local police.

But 2002 was different. The number of people affected over wide geographical areas was enormous. We joined the Citizen's Initiative (CI), a network of NGOs that had been formed the previous year to support relief to the victims of the Kutch earthquake. In the first meeting held at the Gandhi Ashram on 3 March 2002, CI decided to form teams that would do damage and need assessment, assist in legal support, provide food and medical assistance, and coordinate with government agencies. We took the responsibility of procuring food grains for the affected people in the relief camps on behalf of the Citizens Initiative. Other members of the group took the responsibility of distributing these food grains to the camps.

Emergency relief in camps

After our experience of having done similar purchase of food grains for the Kutch Earthquake victims in 2001, we were confident of managing the procurements for the camps. But we were in for a rude shock! We soon found out that due to the communal nature of the riots, food wholesalers, who mainly belonged to the Hindu community, refused to supply foodstuff, even though we were willing to pay market prices. Our staff members decided to contact the stores from where they bought their monthly rations for the bulk procurements. Through this personal networking, we able to supply about 113 tons of foodstuffs, mainly consisting of wheat, rice, cooking oil, potatoes, onions, and milk powder, to almost 37,000 people in 29 relief camps for the initial five days, after which, the government machinery stepped in. Our team worked 24x7, finding and convincing retailers and traders, mostly from the Hindu community to spare a few kilos for the camp inhabitants, to achieve this.

We then developed a short-term strategy to work with refugees in the camps. Subsequently, we worked on long-term rehabilitation of the victims through the Integrated Slum Development programme. We decided that we, and our associate community based organisations (CBOs), Sakhi Mahila Mandal (SMM) in Guptanagar and Ekta Yuvak Mandal (EYM) in Sankalchand Mukhi ni Chali would work with children, women, and youth in the relief camps and provide emotional support.

The managers of the camps initially welcomed us, as they were familiar with Saath and the Citizens Initiative. However, when we introduced them to members of the SMM and EYM, who happened to be mainly from the Hindu community, they were hostile. They said that members of the Hindu Community had caused so much harm, it was difficult to accept that the Hindu members of EYM and SMM would provide helpful emotional and other support to the Muslim inhabitants of the camp. We explained the secular nature of Saath and the CBOs, the type of work we had done with all communities in Ahmedabad since 1989, after which they reluctantly agreed to allow us in the camps. They indicated that they would keep a close watch on our activities and would not hesitate to throw us out, if anything untoward happened. Similarly, the women and youth members of the EYM and SMM were afraid too. They were unsure if they would be accepted in the camps. Further, the husbands and fathers of the CBO members were not at all keen on their wives and children working in the camps. In this mutually suspicious atmosphere, we started working in the relief camps.

Saath, EYM, and SMM worked with the residents of five relief camps at Jamalpur, Behrampura, Ramol, Saraspur, Juhapura for eight months before the camps were disbanded in October 2002.

We found that children and women were the most traumatised residents in the camps. We decided to carry out activities, which would keep them occupied, as well as provide a platform for emotional support and counseling. With support from Save the Children, we started Balghar[10] type of classes for children between the ages of 3 and 10. The teachers of SMM trained women in the camp in basic pedagogy. Professionals from The National Institute for Mental and Neuro-Sciences (NIMHANS) explained the basics of counseling to the teachers, who in turn counseled the residents. Teaching material, games, toys, and stationery were provided. Balghar classes were conducted in

---

[10] Described in Chapter 9, Page 102

the camps for five hours a day and children were given nutritious snacks. Gradually, when the fear and trauma had reduced, trust between camp residents, Saath, SMM and EYM members increased. Almost 1,190 children attended the Balghars in the camps. As part of CI initiatives, we facilitated the provision of food coupons, government assistance, legal aid, and livelihood support for the residents.

The relationships and trust built during the eight months at the relief camps became the foundation for our long-term rehabilitation of the residents.

Long term rehabilitation

We linked development, peace building, and reconciliation in the long-term rehabilitation process; made it part of the Integrated Slum Development Program for riot victims. We selected four clusters of slums in the Saraspur, Berhampura, and Santoshnnagar in the eastern part and, Sankalitnagar and Juhapura in the western part of Ahmedabad with a total of about 2,100 households. Care India, through its Gujarat Harmony Project, and the American India Foundation, provided programme support.

The key to effective ISDP is building capacities of local people and communities to address developmental needs. We identified and recruited women and young men, who were inclined to work towards development and peace, into our team. The new staff members were trained through theoretical inputs and on the job-training, by members of SMM and EYM. They later formed CBOs in their local areas.

The various ISDP interventions in the rehabilitation phase included community health, education, livelihoods, savings and credit, and community participation.

Community health

Our assessment revealed that neurotic conditions such as hypertension, migraine, insomnia, and anxiety related to the riots were prevalent, especially among children. Awareness on maternal and childcare was minimal. Tuberculosis was prevalent.

We trained local women as Community Health Workers. Regular camps were held for immunisation, growth monitoring, and gynecological issues. We started TB referral centers in all the areas and, a DOT center in Juhapura. A dispensary was opened in Juhapura to provide affordable and basic curative services.

People of both communities accessed the various services. Home visits made by Hindu workers to Muslim households and vice versa, which had seemed like an impossibility, was part of the development process that took place. This led to a decrease in prejudices on both sides.

Education

The education initiatives aimed at decreasing the dropout rate of slum children in schools. We started fifteen Balghars for children between the ages of three and five. Picnics and outings, which are an integral part of Balghars, helped tremendously in establishing interest, credibility, trust, and rapport. When the first picnic was held, parents were reluctant to send their children, especially girls, away from home. Later, the parents demanded that more outing be held. The mothers felt that the Balghars had helped the children psychologically get over the trauma of the riots.

Supplementary classes were introduced in Juhapura, Behrampura, and Saraspur to provide additional help to school going children. We started activity centers in the four areas, as interaction platforms for members of all ages in the community. The centers were primarily a library with various indoor games such as chess, carom, badminton, etc., which were an added attraction for the children. Various competitions like hair styling, sports events, etc., were held at regular intervals. The centres were run by local committees, which selected the magazines, books, and games. These participative processes led to a sense of ownership among the residents.

Livelihoods

Tailoring classes, which had begun as a rehabilitation activity in the camps were continued as a six-month certificate course. The classes became a forum for girls to interact (all communities), and discuss various issues regarding adolescence, puberty, etc., which was not possible at home. These classes increased interaction of the two communities at all levels – teachers, students, and parents.

We partnered with AWAG. They had a production centre in Saraspur, which exported embroidery and other specialized tailored items. We trained women for the high quality export work.

Savings and credit

Savings and credit groups were formed to reduce vulnerability by increasing savings and facilitating institutional credit. The response in all the areas was heartening. The initial fears that we would dupe them of their hard-earned money, quickly faded away. The credibility established by us in other programmes played an important role in gaining their trust. But more important, was the fact that the responsibility of the programme was given to the residents from within the community. They were trained to handle the programme. When the residents feared losing their savings, the coordinators assured them that they would take the responsibility of their savings. These groups later became part of the Saath Credit Cooperative Society.

Electrification

In Juhapura, Saath and Sanklap Mitra Mandal, the local CBO, partnered to provide legal electrical connections through the Slum Electrification Programme. After the initial 875 connections during the pilot phase, almost a total of 4,000 connections have been given.

Community participation

The crux of ISDP is capacity building of the people. The process of imparting training to local workers in various programs and making them capable of managing these programs was an enriching learning experience. Members of SMM and EYM, who were mostly Hindus, would be the torchbearers in the Muslim areas where the work was carried out. The resulting interaction between the Hindu and Muslim workers reduced prejudices and preconceived myths. This, in turn, percolated down to the communities. A more positive mindset emerged. Through parents' meetings, street meetings, sports events and various other programmes, people of both communities had the opportunity of engaging on the same forum.

In Juhapura/Sankalitnagar, Sankalp Mitra Mandal was registered in May 2002. Sanklap promoted a volleyball team to engage the interest of the youth. It organized an unprecedented volleyball tournament in Juhapura between thirteen teams from different areas of the city. The non-Muslim players from other teams were of the opinion Juhapura was a 'restricted zone' for them and had never dreamed that they would play a match in the area. The tournament raised hope for the future, helped in partly removing the taboos that existed.

Sankalp established good linkages with the gram panchayats of Maktanpura and Vejalpura. It coordinates with them for various programmes aimed at the development and betterment of people. For example, it helped in the mapping of houses, which were not under the Maktanpura gram panchayat. This helped in those houses getting individual water connections, streetlights, and other amenities.

The long-term rehabilitation work demonstrated that trust could be rebuilt between communities on a development platform within participative processes.

Relief and rehabilitation in infrastructure projects

Infrastructure projects, which acquire land and other assets, lead to disruption of social and economic activities. People affected by infrastructure projects lose land, housing, productive income generation, assets, and community structures. To ensure that losses due to land acquisition and resettlement are minimised, the government of India and international bilateral agencies have recognised the need to resettle and rehabilitate Project Affected Persons (PAPs). NGOs are recognized as sensitive intermediaries between the affected people and government authorities. They are given the task of ensuring PAPs are compensated, rehabilitated, and resettled as per the policies of the government and bilateral agencies.

National Highways Authority of India (NHAI) Project

In February 2003, the National Highways Authority of India (NHAI) entrusted Saath, in collaboration with the CRCB Kassar Trust, with the implementation and Resettlement Action Plan (RAP) for four-lane and upgradation of National Highway 8A & 8B from Samakhiali to Porbander, covering a distance of 311 kilometers. The project was supported by the Asian Development Bank, which had comprehensive guidelines for resettlement and rehabilitation of PAPs.

The broad tasks of the RAP were:

a. Identification, verification, and counseling of Project Affected Persons (PAPs)
b. Educating the PAPs on their right to entitlements and obligations, and ensuring that these were given to the PAPs
c. Preparation and implementation of micro resettlement plans, which included income restoration, livelihood training, alternative housing, and land allocation
d. Disbursing financial assistance to PAPS
e. Assist the PAPs in relocation and rehabilitation, including counseling and coordination with the local authorities
f. Assist the PAPs to address their grievances (through the grievance Redressal cells set up by the project)
g. Impart information to all PAPs about the functional aspects of the various district level committees set up by the project and assist them in benefiting from such institutional mechanism
h. Assist the Project Implementation Unit (PIU) of NHAI in ensuring social responsibilities of the Project, such as, compliance with the labour laws, prohibition of child labour, and gender issues
i. Collect data and submit progress reports for NHAI to monitor progress of the RP implementation
j. Assist PAPs and the CRRO identify and negotiate new land for resettlement

A total of 336 families were rehabilitated with provision of a plot of land and infrastructure like water, and sewage facilities, near Navagam in Rajkot. 1679 families were assisted in procuring compensation.

MEGA (Metro link Express for Gandhinagar and Ahmedabad) Project

Harshad Prajapati, a resident of Vejalpur, had his electric rewinding workshop, at the Mangaldeep Complex at Jivraj Park. Besides the rewinding work, he had tie-ups with wiremen who did other electrical job-work. His father is an auto rickshaw driver. During the planning stage of the Metro, a number of meetings were held in the Jivraj Park area where his workshop was situated. The authorities had initially informed them that there was a possibility of the block of commercial shops/workshops being in the path of the Metro but had not confirmed it.

Harshad's shop was over thirty years old. He had bought it four years ago. It was a small space in the basement of the building. During the monsoons, he would have to physically empty out the rainwater that flowed into his shop. Nonetheless, he was doing good business. His wife is his working partner; she too works hands-on at the shop. "I had a monopoly on this work in the area. I was well known and had a regular clientele. It is an essential service of sorts. People wake up and find their pump isn't working, and cannot pump water into their tanks, they call. It is an emergency," he said.

When the Metro plans were finalised, he said he was happy to find that his shop would be demolished. He had hoped it would, he would then be able to get a better space. Which happened.

When it was finalised, they held a number of meetings with those affected. It was during those meetings that he came in contact with Saath, through Madhviben and Mayankbhai. They were extremely helpful and guided him through the entire process. "I will never forget them for the way they helped me," he said.

"I was asked to find a place to relocate my workshop to. I found the one I wanted. It is in Vejalpur, where I live. In fact, right opposite my residence," he said, smiling. All the shop owners in the building, which was demolished, were given space in the place he had located. It was owned by the AMC. They have given him the relevant documents like the Allotment Letter and Possession Letter and Ownership Certificate; he did not have a Sale Deed. He was told that he would be able to apply for a loan or even sell the premises on the basis of these documents. He had not checked it yet.

"I am now an owner of a much larger workshop, twice the size of my old place. It is newly constructed, about three years old. It is fitted with electricity and water connections. My income has dropped by almost fifty percent. It does not trouble me. It will take me a few years to get to where I was. I will have to build my reputation and clientele, once more. The work with wiremen brings in about thirty percent of my income. It's not enough, but we are able to make ends meet, for the time being," he concluded. His sense of total satisfaction was palpable, as he walked out of the room.

MEGA (Metro link Express for Gandhinagar and Ahmedabad) is constructing a metro rail service between Ahmedabad and Gandhinagar cities covering a distance of 37.7 kilometers. The Japanese International Cooperation Agency is providing financial assistance for MEGA. Saath Livelihood Services was inducted to implement the Resettlement Action Plan (RAP).

As it is an on-going project, according to the Social Impact Assessment (SIA) report, about 1,047 families (3,597 persons) will get affected. Total properties that will be affected are 1,047. Out of these 548 are residential, 351 are commercial (including one industrial unit) and 144 are residential cum commercial units. Further, structures that will be affected due to land requirement for the project have also been identified.

The (RAP) included providing entitlements to the affected families as per the GOI and JICA guidelines, and legal provisions. All PAPs getting directly and indirectly affected are to be provided with their rightful entitlements and compensations, which include alternate housing and new business locations, shifting costs, and restoration of livelihood.

Our work was initiated by an information awareness campaign amongst the PAPs regarding the MEGA resettlement guidelines for providing compensation, assistance, including livelihood restoration opportunities. This included personal and group meetings, as well as public consultations, to help create awareness and build trust between Saath, MEGA, and the PAPs.

In the next phase, we verified persons and families that were identified by a socio-economic survey carried out by MEGA. Persons and families that were left out during the survey were added and identity cards issued to all affected by MEGA.

After identifying all the affected individuals and families, we are ensuring that compensation according to the policy and guidelines of MEGA, which are based on the Land Acquisition Act of 2013, is given to both title and non-title holders. For residential non-title holders, EWS housing measuring 36.5 sq.mt. and the construction cost of their old property is provided. For commercial non-title holders, MEGA tries to provide alternative business locations, which could be a shop or a plot of land. For non-title holders, residential, a shifting allowance of Rs. 50,000 and a subsistence allowance of Rs. 36,000 is provided. Mobile vendors get Rs. 18,000 as financial assistance. Commercial non-title holders receive a one-time financial assistance of Rs. 25,000, a one-time shifting assistance of Rs. 50,000 and construction cost of their old structure.

PAPS, especially the poor and who are from the informal sector, are the most vulnerable in city infratsruture projects. They are not aware of the entitlements that are due to them. We try to ensure that they get their entitlements by not only making them aware, but handhold them diring the whole process, and when they are not satisfied, assist them through the grievance redressal procedure set up by MEGA.

Part 2

CHAPTER 13

ORGANISATION DEVELOPMENT AND GOVERNANCE

This chapter details how we build capacities of individuals and teams at Saath, the values that drive us, governance and leadership transition

Not-for-profit organisations are platforms for social change. To be credible, they have to be effective, collaborative, well governed, and efficiently managed to fulfill its vision of social change. The foundation of any organisation is the values they uphold, practice, and promote.

Building Teams

At Saath, empathy is the fundamental value that drives us – empathy with communities, staff, participants, and partners. For us the core of empathy is being non-judgmental. Our main work is with people who make up communities, that are vulnerable, excluded, and poor; not because they are in any way inferior, butinferior but have been adversely affected by the inequitable nature of India's economic and social structure. They do not have the opportunities for growth that are available to their better-off brethren. However, people in these communities tend to blame themselves for their conditions. Some believe it is their own shortcomings and fate that is the cause of their situation. They tend to believe that there is something inherently wrong with them. In fact, this belief is partly the cause of their poverty.

Our central approach is that every individual has unlimited potential. Our engagement will help unfold latent abilities, which will lead to social change and development. People are central to the work that Saath does.

Ours is a nurturing environment where people get to explore their potential and understand and realise their vast capabilities. This approach has been our template for organisational development, as well as for people and communities we work with.

Our vision is to promote equitable and inclusive societies through empowered individuals and communities. In the process of working towards this goal, the team members are better able to understand the vulnerabilities of the community and the rationale for our Integrated Development methodology.

For operational reasons, our work on the ground is broadly divided into livelihoods, health, education, financial inclusion, and shelter sectors. Various programmes have been designed for each of these sectors based on the needs of communities. Trained teams manage and implement the programmes.

Our workforce comprises of professionals with a sense of purpose, and women and youth from the communities we work with. Those from the community are generally not highly educated or qualified in any particular field. They gradually pick up the ropes from those guiding them and, some join the management cadre within the organisation. Almost 70% of our staff members are from the communities we work in.

Programme designing is a team effort, which includes the community, subject experts, and relevant staff members. Analysis of the situation is carried out and a programme is structured to include financial and human resource requirements and collaborative partners. Methods of implementation, timelines, targets, and review mechanisms are set in place. This process has provided efficient management and effective implementation leading to strong capacity building and empathetic leaders.

The Balghar preschool programme in Guptanagar in 1993–94 is an apt example of our success through this process. We initiated the programme with a socio-economic survey to determine the educational status of children in the area. The findings highlighted the huge drop out rate, especially of girls, in their primary schooling. Talks and discussions with the children, their parents, and the teachers were held.

A supportive learning environment at home was the basic issue. This was due to the fact that most parents had never been to school and were therefore unaware of the need of a learning environment. The expert called in advised setting up of a Montessori education based preschool programme to build a strong and stable foundation for studying. Further advice was to make the parents accountable by educating them on their role in their children's learning process.

A professional recruited for designing and managing the programme laid out the requirements – teachers, rooms for classes, a robust teaching and learning methodology and parental involvement. Rooms were rented out in Guptanagar for classrooms. Four women, with 10th grade education, were recruited from the community to be trained as teachers. The Montessori method chalks out the milestones of a child's cognitive development, as well as records their development. With the help of a child psychologist, the Montessori approach of teaching was fine-tuned to suit the programme.

After being trained in the Montessori method, the teachers visited households and interacted with parents, explaining to them the new method to be used for teaching their children. These visits created a rapport between the teachers and the parents. Recruitment was easier. Subsidised fees were charged creating accountability on both sides. Subsequently, the manager, as well as the teachers, met to discuss the progress of each child. This enabled the teachers to review their own inputs and improve where necessary.

At the programme management level, monthly review meetings were held to monitor various parameters – number of children enrolled, visits to parents, budgets, fees collected, exposure visits, and feedback from the community. These meetings enhanced the management capabilities of both, the teachers and the manager.

An annual review measured the actual programme objectives and targets achieved. In the first year, 70 children had attained measurable cognitive milestones. The teachers had regularly interacted, on a monthly basis, with the parents (of all communities) and managed to keep them convinced of the value of the teaching system introduced.

The manager had transformed four women with a basic education into capable teachers. Through this programme, the manager and the teachers were able to bring out their latent capabilities through the new methods. They became competent, confident, and willing to take on new responsibilities. They had empowered themselves. On the other hand, the parents and the community as a whole, acclaimed the change the teachers had brought about and looked up to them as leaders. By 2005 the first teachers in the Balghars were managing 191 Aanganwadis in eight wards of Ahmedabad with 8,900 children.

Similar capacity building and empowerment took place in other programmes. In the Mother and Child Care programme, local women were trained as Community Health Workers. In the Savings and Credit Programme, community women became Field Officers. In the Youth Programme, the local youth became Social Animators. The Professional managers realised that although staff from the community were not professionally qualified, they were capable of learning and becoming subject experts who could individually manage programmes.

This was possible because we celebrate diversity and gender equality, accept failures, appreciate efficiency, practice collective leadership, and above all, are non-judgmental.

We invite experts on organisation development to assist us with our vision building, long and short-term planning, team building, and creating relevant Management Information Systems. We encourage participation in individual, as well as group self-development workshops.

For management purposes, we work through a hierarchical structure. However, it is not dictatorial. Everyone's contribution is sought and valued. Our annual review and planning events reflect everyone's role and contribution in each programme. Annual plans are finalised after the entire team reach a consensus. Similarly, periodic reviews of all programmes are carried out in a similar fashion.

Collective leadership

Our collective leadership is our Strategic Management Group (SMG), which comprises a mix of senior management staff of middle-class professionals and senior leaders from various communities. They represent organisations and programmes promoted by us. This forum helps uphold the Saath culture and values. It maintains a balance between the community needs and the programme/donor/partner, as well as facilitates linkages between the various programmes and entities. A major role of the SMG is to take important decisions that affect Saath's growth. The SMG revisits Saath's vision, mission and objectives every five years, based on the needs of the communities we work with. This has enabled us to remain relevant to the community and all the stakeholders. Programmes are then aligned to the changed objectives.

Students and interns from all over the world are welcomed and encouraged to work at Saath. Engagement between interns, staff members, and communities has been transforming. Interns learn to appreciate the way vulnerable communities live normal lives with limited resources. With their varied exposures, they bring fresh perspectives in to the different programmes. Staff members and communities are enthused when "outsiders" appreciate the social change taking place. Most of the interns, after interacting with vulnerable communities, have commented, they will "never want for anything in their lives at home."

Governance

Saath has gone through the typical governance cycle of not-for-profit-organisations. During the first twelve years from 1989 to 2001, we had a board mostly comprising of the founder, senior team members, and a reputed well wisher. During this phase, the founder made the key decisions, as well as took care of governance covered regulatory compliances, proper accounting systems, and regularity of reports to funders. In 2001, when we started working in Kutch after the earthquake, the amount of annual funding received almost tripled. We needed to revamp our accounting system to ensure that the funds were properly disbursed and accounted for. Once again, in 2002, after the Gujarat riots, we were actively involved in relief and rehabilitation of the victims, which involved large volumes of funds and material. There would be much more scrutiny of Saath from the regulators, donors, as well as the government. We needed a more robust governance system.

In 2002, we requested Mr. Gagan Sethi, a well wisher of Saath since its inception, and an expert in board governance in not-for-profits, to create robust governance systems and to guide Saath's transition from a founder-led organisation to a board-led organisation. The first and essential change was the founder stepping down from his position of independent charge and relegating decision-making to the forum. From the founder's perspective, this was a critical and risky decision. However, Mr. Gagan Sethi's mentoring helped in making the shift, since he had made a similar change in Janvikas, the organisation he had founded. Over time, the founder of Saath realised that an effective and supportive board empowered him to lead more effectively.

The guidelines for becoming board members changed, staff members, and relatives could not be members. Reputed persons from the development and corporate sectors, judiciary, academic institutions, and bureaucracy have been members of the Saath Board of Trustees. The board appointed a governance officer who assisted the Executive Director in all governance matters and was accountable to the board. Policies for audit, procurement, etc. were formulated and approved by the board. The board formed a finance committee, which would be responsible for all finance and fiduciary related matters. The board created a rotation policy for the trustees. We can confidently claim that board governance has made Saath transparent and accountable to all its stakeholders, which are values that we cherish.

In 2005, we became members of the Credibility Alliance, a consortium of voluntary organisations, committed towards enhancing accountability and transparency in the voluntary sector through good governance. Presently we are accredited with meeting the "Desirable Norms," the highest rating.

To meet the needs of communities and ensuring program focus, Saath has promoted various entities. In 2007, Saath Livelihood Services was incorporated as a not-for-profit Section 25 company to focus on livelihood interventions. In 2010, the Saath Savings and

Credit Cooperative Society was registered to concentrate on financial inclusion. Similarly, the Saath Mahila Savings and Credit Cooperative was registered to meet the needs of rural communities that we worked with. In 2017, a for-profit company, Saath Facilitators Pvt. Ltd. was formed to take forward the revenue generating social enterprises started by Saath. In 2018, Saath Grameen was registered, as a trust to address rural needs.

Leadership change and succession

In 2009, the founder felt that Saath needed a different leadership to guide it through the changing times. A new leadership would bring in fresh energy, innovative programme formulation, and new ways of management. At a personal level, he felt that he was stagnating and needed new challenges. He wanted to explore new forms of collaboration between the for-profit and not-for-profit sectors to promote social entrepreneurship. He shared his concerns with the Saath board and the senior team members. There was apprehension that a new leadership would change the organisation drastically. We decided that a consultative and transparent succession process would assuage the fears. The board asked Mr. Gagan Sethi to facilitate the succession and, a new leadership emerged for Saath. The founder, however, continues to be a mentor to Saath and the various entities that have been promoted by Saath.

Entities Promoted by Saath and Meta Governance

All these entities have their separate boards, leadership, and governance systems. At the same time, they use the Saath brand for credibility, outreach, and growth. The parent Saath Charitable Trust board has instituted a meta-governance method to ensure synergies between the organisations, common governance principles, and avoiding conflict of interest.

Saath's Leadership Transition

In 2009, Rajendra Joshi, founder of Saath, informed the Saath Board that he would like to step down. Having founded Saath in February

1987, he had worked for 20 years at the helm and now felt the need for a personal and professional shift. He had started out as a field worker and, over the years transitioned into an internationally recognized expert on urban development. He felt the need to utilize and share his experience at a wider forum. He wanted to take forward his ideas and vision beyond the "not-for-profit" space.

At the organisational level, Saath was widely recognized as a credible organisation that had created tremendous impact on the urban development landscape. It was in a phase of growth, and there were many young mid-level managers in the organisation. The Board of Trustees was strong, diverse and very involved. An Executive Committee for financial affairs was actively working to support the organisation through any financial crisis. Most significantly, the communities and their aspirations and needs were changing.

At the end of 2009, Mr. Joshi finalized his decision on moving on after identifying new leadership and ensuring a smooth transition process. The Board decided to appoint an external facilitator, Mr. Gagan Sethi, an ex-trustee and mentor for Saath.

Mr. Sethi, an Organisational Development expert agreed to facilitate the entire process. The founder and consultant had many preliminary meetings to clarify the expectations of the founder and then finalised the process to be followed to appoint the new Executive Director.

The fact that the facilitator had been an ex-trustee and mentor to the organisation helped people in Saath to accept him more readily. The facilitator's mandate was made clear to the team and they accepted his role and authority in the matter.

This process also required that the founder, at some points, be purposefully kept at a distance – especially when the leadership and capacity building process was undertaken in the first year.

Sharing the Plan Internally

The first step was to share the plan. At Saath, this was done at two levels. The first was with a very senior manager and mentor, who most people had anticipated would be the next in line, and the second, was with senior community leaders and program managers across the organisation – people who were part of the core of the organisation. Bringing the senior manager into the process was crucial; she was the link between the new and old. Her support was necessary to build momentum for the shift. This dialogue brought about an openness to the process, as well as clarity and assurance that nothing was being done for personal gain, but rather for the continuity and sustainability of the organisation.

The team was very supportive even though they had doubts and concerns about the founder not being a part of Saath. The founder's role was going to change, but that would not mean that he would leave Saath – it was a shift in the role, more than a disconnect. It would be important for the new team to carve out his role at a later stage, which they understood and accepted. Community Leaders who had been with the organisation since it's inception were the most important stakeholders. They needed to believe in the need for change in leadership. As they had been brought on board at the very beginning, they were honest critics at every point. They were the supporters of this change that convinced the rest of the organisation, right down to the field workers.

The Process and Mechanism of Selection and Transition

Saath had gone through a process of hiring young, new talent over the years – these were young professionals from different backgrounds and orientations who were managing and handling programs. Eleven of them, in the age group of 27–35 years, were the ones selected and invited to join the process. Being a 20-year old organisation, there were many senior leaders at program and community level. However, the rationale for this was, that a younger team would bring in fresh ideas. A younger team may also have a longer shelf life to take the organisation ahead.

At the end of the year, two people were selected – Ms. Keren Nazareth and Mr. Niraj Jani. They were made Co-Directors and were given different programs and aspects to handle for the organisation for a period of a year. Mr. Joshi and Mr. Sethi, continued to mentor them.

The next step was another year of divided responsibilities between Keren and Niraj, including organisational responsibilities – HR,

Governance, Finance, Programmes etc. It was a year of working directly with the founder and taking decisions along with him for the organisation, as well as being involved in Governance and finance. What was unique about this year was that during the process of leadership development, the core team was formalised and a Strategic Management Group (SMG) was set up as the evaluating body for them. This SMG became the conscience keeper of the organisation and held each member of the team accountable for the work as well as for upholding the values of the organisation. Both the Co-Directors were answerable to the SMG. The role of the SMG was to provide feedback of the working of the Co-Directors and their feedback was crucial to the final selection.

The year culminated with Keren Nazareth being selected as the Executive Director. The original thought had been of only an ED, but considering Niraj's expertise, contribution and passion for the work and the organisation – the idea of a joint leadership emerged. Niraj was offered the position of Associate Director. Thus, a model of joint leadership was built to take Saath forward.

The model of joint leadership held Saath in good stead in the subsequent years, because the two people who were leading the organisation had different strengths and supported each other through difficult periods.

Rajendra Joshi's role changed, he became a mentor and consultant to the organisation, as he continues to be on the Board at Saath.

In December 2014, Keren Nazareth informed the Board that she would like to step down from the position of Executive Director, as she had received an opportunity to work in the space of animal welfare, which had been her lifelong dream. She and Niraj underwent a period of transition for 8 months after which, she stepped down in September 2015.

Niraj Jani, took on the mantle of Executive Director and led Saath till October 2018

Collaborations are integral to Saath. The next chapter describes how we have partnered with stakeholders

CHAPTER 14

PARTNERSHIPS

Organisations that want to make significant social impact cannot work in isolation because the issues that they address are a result of structural fault-lines in policies, neglect by the state, ineffective implementation of government social programmes, and societal indifference. In working towards our vision of creating inclusive societies, we have understood the importance of working in collaboration with relevant stakeholders. The word "Saath" means 'cooperation and working together,' in Hindi, as well as in Gujarati. Working in partnership is in our DNA. Our partners have been the communities that we have worked with, municipal, and state governments, donors, academic institutions, NGOs, private sector companies, professionals, and Corporate Social Responsibility initiatives.

We have inculcated the principle of constructive partnerships in all our programmes. For us, genuine partnership means trust, co-creating programmes, sharing resources and information, joint reviews, sharing credit and recognition, and respecting different objectives. People represent partners we respect, trust, and accept. Overall, our partnerships have been successful and long-standing.

Communities

In all our work with communities, they are our primary partner. We initiate our work with them through their various caste, political, economic, and social leaders. For deeper understanding of the local issues of the community we work with the people from the community. We discuss our vision and objective with them, discuss their issues and patiently answer their questions. We carry out these discussions at a place where they are comfortable. To help us understand their socio economic issues and needs, we recruit local community men and women. Survey and needs assessment tools are jointly designed and presented to the local leaders for approval. Results are shared and a jointly designed plan is finalised. Finally, a committee comprising of community members and Saath staff is formed to oversee the work is carried out as per the mutually sanctioned plan. Working closely with the people directly inspires trust and confidence, the foundation for any meaningful work. Financial budgets are open for scrutiny by the community.

Donor Organisations

We have been singularly fortunate with the support that we have received from development-oriented funders. When we began work in 1989 without any track-record, Oxfam and the Indo-German Social Service Society put their trust in us and provided us with our first seed funding. Cordaid, a Dutch based funder, unstintingly supported our Integrated Slum Development Programme between 1993 and 2002. They were our principal funders for the relief and rehabilitation work during the Gujarat Earthquake. We have these three agencies to sincerely thank for helping build a strong foundation for Saath. They trusted in our mission, commitment, and us unconditionally. They provided resources for institutional growth as well.

CARE international and Action Aid funded our work after the Gujarat riots in 2002. The American India Foundation believed in our innovations in livelihoods training by providing seed and scaling up funding.

Save the Children were our partners for eradicating child labour in the cotton growing areas of Ahmedabad district.

The Ford Foundation is supporting our work of facilitating housing for vulnerable migrants. Their support is somewhat unique, in that it includes building the organisation without emphasis on targets and timelines.

In 2005, with Counterpart International, a USA based organisation, and the Ahmedabad Municipal Corporation, we scaled up our mother and child care programme through the Jeevan Daan Maternal and Child Survival programme.

Grant support for programmes have also been received from Action Aid, American India Foundation, ASHA Education, Ashoka, CARE International, Cordiad, Counterpart International, Dell Foundation, Empower Foundation, Ford Foundation, Give Foundation, Global Giving, Indo German Social Service, IPartner, Mobile Creche, Oxfam, Paul Hamlyn Foundation, Quest Alliance, Shivia Foundation, UNICEF and Vibha.

Municipal and State Governments

Basic services of sanitation, water supply, and drainage, in all cities, are taken care of by Municipal Corporations. Our association with the Ahmedabad Municipal Corporation (AMC) goes back to 1989, when we worked with the residents of Sankalchand Mukhi ni Chali to build toilets under their 80:20 scheme. To effectively carry out our Integrated Slum Development, we partnered with the AMC. In 1995, we jointly created the Slum Networking Programme with them; a path-breaking initiative, in which, for the first time, Communities, the AMC, and an NGO (Saath) shared a complete partnership. Financial costs for providing basic services of sanitation, water supply, drainage, paved roads, and streetlights were shared. The project won national, as well as international accolades and, recognition of all the partners. Being an intermediary between the community and AMC during the pilot projects in Guptanagar and Sanjaynagar helped hone our skills in partnership relationships. The trust between Saath and AMC endures, though we have differences on policy formulation.

We partnered with the AMC again for piloting Udaan, a youth employability training programme in 2007. In this collaboration, AMC provided space and paid costs; Saath mobilised, trained, and made placements. This programme, renamed Umeed, was replicated across the state covering almost a lakh youth between 2009 and 2012.

Our partnership with the state government during Umeed had its ups and downs. The good part was working with senior bureaucrats in designing and implementing a state-wide programme. We got to understand the intricacies of the working of the bureaucracy and the tremendous power it wields to do good. We also worked with the Women and Child Development department managed almost 180 Aanganwadis in Ahmedabad.

We have vast experience working with the government. Bureaucrats, the operational face of the government, are powerful people who can make changes in policies and procedures at will. We have had first-hand experience in the matter. For our Umeed programme, the procedures for payment to Saath were changed ad-hoc; we suffered a loss of almost one crore. On the flip side, we have also had the opportunity of working with some very committed government officers.

Another drawback with government welfare programmes is that they have cut and dried schemes, with no place for flexibility. NGOs are more or less in the role of contractors to the schemes, expected to follow the norms set, exactly. Our Balghars were run with the premise of accountability all round. Affordable fees were charged. When we suggested the same for the Balghars being run through the government Integrated Child Development Scheme, it was shot down. We withdrew from the programme.

Private Sector Companies

Historically, NGOs and the private sector have had an adversarial relationship. Differing ideologies, a perception that the private sector is exploitative, that NGOs are not in sync with market realities, and the dissimilar working styles are the main reasons. There is a grain of truth in these perceptions, but it would be a mistake to make such generalisations about all companies and NGOs.

Saath has worked closely with companies. In 2000, we partnered with Torrent Power (then known as Ahmedabad Electricity Company) to provide affordable electricity in slums. As the sole utility providing electricity in Ahmedabad, Torrent Power did not consider slum households to be a significant market. This was because the high connection costs (approximately Rs. 10,000) were not affordable for slum households. As legal electricity was not available, slum households had to rely on informal electricity supply provided by the local mafia. Saath and the USAID approached Torrent Power with a pilot project to provide connections to 1,000 slum households for Rs. 2,000 each. We wanted to demonstrate that this was an affordable rate and that the subsequent bills would also be affordable and that slum households would pay their dues regularly. This turned out to be the case and Torrent Power asked Saath to do a larger pilot in which about 5,000 households in the Juhapura area of Ahmedabad at connection cost to Rs. 4,000. The success of the second pilot convinced Torrent Power that slum households were a viable market. The company then reduced the household connection cost to Rs. 2,000 and over the years saw a significant rise in revenues as almost 90% of slum households were electrified. This was a partnership through which slum residents got electricity at affordable costs, Torrent Power increased its market and Saath's objectives of enhancing the quality of life of the urban poor were met.

Our second partnership started in 2010 with DBS Affordable Housing, a real estate company wanting to build affordable houses for the lower middle class. Till 2010, formal housing was beyond the reach of the informal sector, but policy changes by the National Housing Bank enabled people without formal salary certificates to avail of housing loans. With DBS, we created a business model in which DBS, as a developer, would build houses, housing finance companies would provide loans, and Saath would identify customers.

We have worked with Micro Housing Finance Company (MHFC), the first housing finance company to be approved by the National Housing Bank to provide housing loans to the informal sector. With MHFC, we have created loan products for incremental housing.

In these three partnerships, Saath and the companies have co-created products and working business models, which are win-win for the partners and customers from low-income neighbourhoods.

NGOs and Civil Society

The not-for-profit has a lot to gain through collaborations. Partnerships advocating a cause, responding to disasters, and providing services have a far better outreach. As part of a vulnerable sector, with similar objectives of empowerment, and multiple synergies, it does not make sense for NGOS to be competitive.

In 2001, Saath joined the Kutch Nav Nirman Abhiyan (KNNA) a network of NGOs in Kutch working towards the relief and rehabilitation of communities affected by the Gujarat Earthquake. Similarly, in 2002, we joined the Citizen's Initiative, a network of NGOs for the relief and rehabilitation of the Gujarat riots. Our experience working as part of these networks was very positive because resources were used more efficiently, the geographical reach increased considerably, and the expertise and competencies of the partners was complementary.

We partnered with the Mahila Housing Trust during the Slum Networking and Slum Electrification projects.

In the Umeed, Youth Force, and Saloni programmes we worked with many NGOS in other towns to increase the outreach.

Our learning in NGO partnerships is that the goals, growth phase, working style, and competencies have to be respected. Building capacities of partner NGOs has to be done sensitively. A sure deal-breaker is an attitude of patronage and one-upmanship.

Academic Institutions

In 2005, Saath partnered with Centre for Health Management at the Indian Institute of Management, Ahmedabad, to design and establish a model Urban Health Centre with focus on people living in slums and chawls. The model was based on extensive use of Geographic Information System (GIS) to locate urban health centres in the Vasna, Sarkhej, and Paldi wards, to ensure availability, access, affordability, and equity of healthcare services, and as a viable alternative to private health care providers.

We have partnered with professors and students at the Centre for Environment Planning and Technology (CEPT) for research on issues of urban poverty and, have co-authored articles.

With Sangath, we have hosted a studio for international students on the issues of slum housing.

Students from Duke University, USA and Utrecht University, Netherlands, have interned at Saath under formal agreements. We have carried out joint research with Mahatma Gandhi Institute and Abdul Latif Jameel Poverty Action Lab (J-PAL).

Corporate Social Responsibility Partners

In April 2014, the Indian government amended The Company Act to make Corporate Social Responsibility (CSR) mandatory for companies with a net worth of 500 Cr or more, or a turnover of 1,000 Cr or more or a net profit of 5Cr or more. The companies have to spend at least 2 percent of their net profit for the preceding three years on the national development agenda. Companies can partner with not-for-profit entities for carrying out their CSR objectives.

At Saath, we have engaged extensively with CSR activities of almost seventeen companies.

With Dewan Housing and Finance Ltd (DHFL), Saath has co-created a programme to promote financial inclusion leading to better livelihoods, and transition from informal to formal housing. Participants are engaged in the informal sector with a household income of Rs 2–3 lakhs per annum. The components of the programme included awareness about basic financial literacy, opening and operating bank accounts, savings, investments, and insurance. This CSR activity facilitates Mudra loans, availing housing loans under the Pradhan Mantri Awas Yojana, and skilling. After starting in Jaipur, the programme has been scaled up in Varanasi, Ranchi, and Raipur.

We are partnering with Godrej Ltd in its CSR activity called the "Core Beauty & Wellness Program." The program is built around an intensive and comprehensive training module focusing on core beauty skills and entry level beautician training for women. Components of the program include, self-awareness, health & hygiene, women's rights, and exposure workshops. The partnership started in 2013 when Godrej supplied consumables in the beautician training program of Saath.

In 2014, Godrej started supporting skilling of master trainers and participants. In 2015, Saath became a nodal agency and is now working with 17 partner organisations across Gujarat.

Our CSR partners for girl's education have been Nivea, Intas Pharmaceuticals, and Radio City. For access to water and sanitation, Coca Cola, Ford Motors, and Vestas have been our partners. For youth employability training our partners have been BOSCH, CEAT, H T Parekh Foundation/HDFC Bank, Gruh Finance, HSBC Bank, Tata Motors, IGATE, Microsoft, Ambuja Cements and PWC. KPMG has contributed towards children's education

The social impact that Saath has made over the last thirty years is elaborated in the next chapter.

CHAPTER 15

SOCIAL IMPACT

Social impact can be measured qualitatively and quantitatively. The parameters for measurement can be the number of people who have benefitted, awareness, attitudinal change and, social and economic progress. The effects can be short as well as long term.

Quantitative Impact

The quantitative impact of our work is given in the following table. We have covered 3,80,106 persons directly. These are numbers that have been culled from our robust Management Information System, project reports and our annual reports. We can confidently claim that indirectly we have impacted their immediate families. Considering an average family size of five persons, the number of persons would total to 14,53,255.

Number of participants directly impacted between 1989 and 2018

No	Social Impact Area	Type of Social Impact	Number of persons/Households reached
1	Pre-school Children	Nutrition support, education,	14,468 children
2	Mother and Child Care	Immunization, Safe delivery of Children, Nutrition Support	1,06,321 women
3	Tuberculosis Prevention	Facilitating treatment and nutrition support	1,342 persons cured
4	Children	Reducing Child Labour	9,005 children
5	Education of Girls	Facilitating financial support	725 girls
6	Youth	Motivation, Self-Growth and good citizenship	9,305 youth

Contd...

No	Social Impact Area	Type of Social Impact	Number of persons/Households reached
7	Employment	Motivating, training and placement for jobs	59,613 youth
8	Employment (Disabled persons)	Motivating, training and placement for jobs	499 persons
9	Livelihoods (Home Managers)	Motivating, training and placement	714 women
10	Urban Livelihoods (entrepreneurship)	Motivating, training and handholding	387 persons
11	Rural Livelihoods (entrepreneurship)	Motivating, training and handholding	1,550 persons
12	Incremental Housing	Facilitating basic services, permissions and loans	2,700 Households
13	New Affordable Housing	Facilitating purchase of houses	1,187 Households
14	New Housing during Kutch Earthquake	Monetary support, Facilitating permissions and construction	376 Households
15	Electricity Supply	Facilitating legal electricity	6,500 Households
16	Financial Inclusion	Awareness, facilitating savings and loans	44,680 members, 30,215 members took loans
17	Water and Sanitation	Enabling access, construction of toilets, providing RO water machines	3,780 children
18	Citizenship	Facilitating documentation	38,262 persons
19	Relief and Rehabilitation	Relief and rehabilitation during Gujarat earthquake and riots	7,128 households

a. Persons directly reached: 2,90,651
b. Impact of directly reached on household of 5 persons: 14,53,255 (a x 5)
c. Households directly reached 17,891 (89,455 persons @ 5 persons per household)
d. Total directly reached (Persons + Households): 15,52,670

Qualitative Impact
Children
Immunisations and nutrition support to children has been a tremendous booster to the general and cognitive well-being of the children in our programmes. There have been fewer illnesses and, they have been able to do well in their studies.

Women
Women's empowerment has been through a number of actions in our projects –understanding the value of nutrition, importance of hospital deliveries, and timely immunizations. These changes have led to improved livelihoods with better incomes. The overall impact has been their ability to take better care of their families and a change in their world view. Many are now leaders in their communities.

Youth
The youth have been at a distinct advantage. They were given the right motivation, which enabled them to become self-assured young men and women. They became role models and leaders in their communities; confident and able to deal with city authorizes and other stakeholders.

Communities
The communities have grown from dependent to independent, well organized groups. Their dependence on politicians, who considered them their vote bank, is now non-existent. They are self-assured and feel more secure, their constant fear of eviction is fading. They no longer look for handouts but are able to work for what they need or want.

Electricity, water, sanitation and housing
Access to basic services has enhanced the physical quality of life with a multiplier effect. With sanitation in place, expenditure on medical expenses were reduced, leaving them with more money to spend on educating their children. Women were relieved of the task of standing in

queues for pots of water, giving them more time for their families. With improved housing the environment no longer represents the shabby slum it once was. Being part of this improved environment has given them a sense of pride and achievement.

Livelihoods

Apart from the quantitative increase in incomes, for those who rose from being unemployable to being able to earn and contribute towards the wellbeing of the family has had a huge psychological impact. The sense of achievement has led to the rise of self-esteem and respect among peers and society. They now are contributors to the national economy.

Financial inclusion

Financial inclusion is the backbone of economic empowerment. The ability to bank, save, avail of credit at reasonable rates, get insured has contributed to the social and economic stability of individuals and families.

Quality of life

The people and communities we have served have seen a considerable improvement in the quality of their lives. Their physical surroundings are better and safer. They are healthier and better educated. With increased incomes they are able to save and access loans, even for housing; become owners of household appliances and vehicles as well.

Citizenship

The people we have worked with do not any longer feel alienated; they are equal citizens. They now have documents with which they can access their rights and benefits as citizens. They can engage with the government, middle class, private sector, and civil society institutions confidently.

Disaster, Displacement and Relief

For people affected by disaster and displacement to be able to recover, caring and support is as important as material requirements. Caring heals and smoothens the path to total recovery.

Macro-Impact

When low-income neighbourhoods are transformed, the city and country benefits. Economic productivity and social cohesion increase. Infrastructure and housing get better. People, who felt excluded, participate in national and societal objectives.

Policy advocacy and replication

A number of innovations, programmes, and projects developed by Saath have been replicated. The Ahmedabad Municipal Corporation at the city level took up the Slum Networking Project. The Slum Electrification Project was scaled up by Torrent Power in the cities of Ahmedabad and Surat. Youth employability training was scaled up across the state by the Gujarat government. Other foundations and NGOs have replicated this model. Urban Resource Centres became part of the National Urban Renewal Mission. Social enterprises such as Urmila and Griha Pravesh have been replicated by other organizations. These indicate a much wider social impact of Saath's work.

Recognition and Awards

- Inspired Awards by Brit World Wide and Parivartan in the field of Women Empowerment and Child Welfare 2018–19
- Certificate of Appreciation under HUDCO Award for Best Practices to 'Improve the Living Environment 2017–18'
- Saath awarded Champion Level Platinum Seal by GuideStar India Transparency Awards – 2016, 2017 & 2018
- Saath Annual Report 2014–15 selected among 100 Meritious Publications by ICE Awards, an initiative by Shailja Nair Foundation

- India NGO Award 2014–15 in the Medium Category by the Resource Alliance and Rockefeller Foundation
- Citi Micro Enterprise Award 2013 in the category of 'Innovative Livelihood Promoter of the Year'
- India NGO award, 2011 and 2010 for Western Region by the Resource Alliance and Rockefeller Foundation
- Edelgive Social Innovation Honors 2011 for the Urmila Home Manager Program
- Accenture South Asian Network's Charity of the Year 2010–11
- Indian Social Entrepreneur 2009 by Schwab Foundation, UNDP and CII
- Udaan supported by Microsoft (India) Corporation Pvt. Ltd. awarded E-Rajasthan Awards 2009, Digital Learning – Private Sector Initiative of the Year
- Recognised and Profiled in 2009 by CII as one of the 50 NGOs in Gujarat to collaborate with
- Ashoka Fellowship, September, 2008.
- Listed amongst 50 "Pioneers of Change" by India Today in July 2008
- Karamveer Puraskar for "Real Wealth Creators for Communities" by ICONGO in 2018
- Finalist for Social Entrepreneur of the Year Award – 2007 Constituted by UNDP, CII, Schwab and Khemka Foundation
- Awarded The Nagrikta Puraskar in 2004 by the Ahmedabad Management Association

CHAPTER 16

People at Saath

People nurture and build organisations. At Saath, we have been fortunate to have people who have taken up ownership and have been deeply committed its values and vision. They are from a diverse variety of backgrounds in terms of gender, caste, religion and geographical origin. They are a mix of professionals and community leaders. We have profiled a few below.

Chinmayi Desai, Urban Programmes Director

After my disappointment in my job at the BM Institute of Mental Health, my guide, Mr. Suresh Majumdar, an education consultant with Saath, told me of a vacancy in the education programme that Saath was running in slums. He suggested I should go and meet Rajendrabhai. I had no idea as to what a "slum" was. I had never even heard of NGOs. I didn't know what they did – I had no knowledge of the Social Sector at all. I met Rajendrabhai, after discussions he invited me to join them.

After I spent time observing what the programme was about and how it was run, I was convinced that this was what I wanted to do. I joined Saath in 1995.

My journey started at the Guptanagar-Pravinnagar slum. I worked in the education programme, but at that time there was no clear demarcation. I was also involved with the community in other development spheres. I attended meetings and learned as I went along. I realized that work in a slum was never completed, never over. It was an ongoing process. I realized that this was a long-term process, with the individual, family, as well as the community.

Dealing with day-to-day issues, there were times when I felt a bit weary. However, watching the progress and results, I was glad to be part of the process. I had learned a lot from those already on the job, especially how to deal with everyone and their problems in the right manner. There were people from diverse backgrounds, immigrants from different states too. I was able to observe their efforts at improving their lives.

I learned to plan. Weekly meetings were held, where, after reflections and brain storming sessions, plans were worked out – short-term and long-term. I understood the importance of planning. It brought a lot of clarity into how future tasks were to be carried out, responsibilities divided, support requirements, as well as community involvement at all levels.

I worked for around two years with the education programme, as a Programme Coordinator. As was the norm in Saath, during personal evaluations, one could opt to work in another programme. The education programme was replicated in other areas, and I took over as a Sector Coordinator. Gradually, I was undertaking multiple responsibilities, which included health programmes, and creating and registering Community Based Organisations (CBOs).

After approximately 8–10 years I was appointed on the Saath Board of Trustees for a term of five years. The idea of having rotating members on the Trust Board had been implemented. This was to enable those working in projects understand the working of the Board, bring about a transparency that could be explained to anyone wanting to understand it. This was later withdrawn and the Board reverted to having only external members.

When Rajendrabhai decided he wanted to step down and bring in another person to head the organisation, it initially evoked a feeling within the organisation that everything would start falling apart. We had numerous discussions and, I played a part in smoothening the process. I worked at keeping a balance and restoring trust within the organisation. Explaining the positivity of new direction and decisions went a long way in the smooth transition.

I was also involved in the 2002 riot rehabilitation programme across Ahmedabad. I had now become Coordinator Head and oversaw multiple programmes. I went on to become the Chairperson of both the Saath urban and rural savings and credit cooperatives.

Today, with abundant experience, I overlook multiple programmes and am able to guide, motivate, and direct programme personnel, as well as coordinate with funders.

Often, professionals in the commercial sector seem to fade away without feeling a sense of achievement and satisfaction; they feel worn out by the end of their careers. The NGO sector provides multiple roles and cross-functional workspaces, as well as freedom to innovate. The challenge facing NGOs is finding people with a sense of commitment.

My message to professionals is that just running programmes and managing day-to-day implementation is not what the NGO sector needs. Professionals are required to come in with new, fresh ideas, outlook, as well as bring in different, new technologies.

As a person, I feel my need to reach out to people has been fulfilled. The feeling of satisfaction is tremendous. Looking back, I don't have a single moment of regret. I have leant so much. Working in Saath is an ever-enriching journey I am happy and proud to be a part of.

Ms Devuben Parmar

I belong to a small village called Sapar near Rajkot city. After a huge loss in my husband's business in 1988, we moved to Ahmedabad. After a couple of horrendous accommodations, we rented a hut in Pravinnagar-Guptanagar (PG). This settlement was also filthy, without proper amenities.

Sonalben from Saath visited our neighbourhood regarding a vaccination camp. I took my daughter to the camp where I learned they were about to start a Balghar (pre-school/nursery) in the slum. I have studied up to the 9th grade. I also enjoy working with children and was interested in working at the Balghar. I was one of the two women from my area to be hired. We were first trained in various methods, techniques and also the theoretical aspects of childcare and development. My dedication and hard work made me popular among the children. Gradually parents began trusting me as well.

Discussions on the Slum Networking Project (SNP) had begun in our area. I was part of the Saath team that visited each household, held discussions with them, as well as helped with community meetings. It was a new experience for me; I interacted with people from different communities, irrespective of caste, class, or religion. People began recognising me within the community. I got to know everyone on a first name basis; a very fulfilling experience.

I was appointed Manager of the Sakhi Mahila Mandal (SMM), when it was set up in 1996. After the 2001 earthquake, I collected material for the affected people in Kutch and subsequently went to Kutch. Saath expanded its work in rural areas.

During the communal violence of 2002, Saath initially collected relief materials from the households and when Rajendrabhai asked who among the Saath team was willing to work in the relief camp, I volunteered. All of us knew that there was no compulsion whatsoever, but most of us joined in. A van picked us up and dropped us back at the settlement. I had not told my husband about the work, but he suspected, and later I had to tell him. But I shared my experience of the camps; the sufferings of the victims, how the children felt threatened... I also told him that it was safe to go there.

Working at the relief camps changed my life – it opened my eyes. I too had gone in with fear and prejudice to some extent, but all that changed the moment I stepped into the camp. The stories of horror shook me completely. We started activities with the children and women. Not many people in my community understood why I was doing it, and it was a struggle having to explain myself all the time. But, I did, I shared my experience. Slowly people understood the gravity of what had occurred. Over a period of time, Saath set up an office in Juhapura (2002 riot-affected Muslim area) and today all our programmes are running in Juhapura.

In the year 2007, I was appointed Coordinator of the Urban Resource Centre (URC) for the whole Vasna area.

Authors note: After 2003, Devuben managed 180 anganwadis in ten wards of Ahmedabad and is currently leading the Home Managers programme.

(This case study had been co-authored by Rajendra Joshi for Working Paper 6 'From Basic Services Delivery to Policy Advocacy – Community Mobilisation in Pravinagar-Guptnagar by the Centre for Urban Equity, CEPT, Ahmedabad in April 2010)

Bharat Chauhan

Mr Bharat Chauhan's impassive countenance belies an impressive understanding of community participation and development. He started the interview by stating, "My journey with Saath started 28 years ago when Rajendrabhai and Robertbhai visited Sankalchand ni Chali (SMC) and interacted with us. They were interested in finding out problems faced by the residents. This was the first time anyone inquired about the issues we faced and, I decided to help them in whatever way I could."

Now 50 years old, when Saath started work in SMC in 1989, Bharatbhai was a strapping young man studying in the 12th standard. His grandfather had come to live in SMC 70 years ago.

He was part of the first volleyball youth group of SMC. "After playing volleyball, we discussed our problems. I participated in the first youth camp at Sugadh. I realised that Saath did not have a political, religious, or vested agenda," he said. He volunteered when Saath was conducted a socio-economic survey of SMC. "While doing the survey and later understanding the findings, I realized the importance of data," he reflected.

Bharatbhai led the toilet construction programme in SMC[11]. "I remember meeting the officials at the Ahmedabad Municipal Corporation (AMC) to understand the 80:20 scheme, which I subsequently explained to the residents. We decided that households with sick non-ambulatory, and aged people should be given priority in the first batch," he said. He took care of the entire procedure – right from making the required application with the supporting documents to the AMC and obtaining work orders, to ensuring the toilets were built

---

[11] Described in Chapter 1, Page 20

as per the scheme specifications, and finally obtaining the completion certificates for the AMC. After release of the payment, he ensured that the loans from Saath were repaid. He then repeated the procedure for the next batch. "Before the toilet construction programme, very few people knew me in SMC. After the first batch, residents recognised and respected me," he said proudly, "I became self-confident and learnt how to interact with the AMC officials, residents, contractors, and the bank staff."

When Saath started the TB control programme in SMC[12] Bharatbhai was in the thick of things. Recalling his experience, he stated, "I met TB patients and the doctors at the TB hospital to understand why patients were not given adequate treatment. Very often, patients stopped taking the medication and did not complete the treatment. We found out that the doctors were being rude and insulting, which put the patients off their treatment." As required by the TB control programme, Bharatbhai made sure the patients took their medication regularly, distributed nutrition supplements, weighed the patients to check weight change as well as kept a watch on the regular follow-up with the doctors. "Gradually, people with TB symptoms started asking for help. TB was no longer a taboo topic. I organised TB awareness camps," Bharatbhai said. Ekta Yuvak Mandal helped TB patients get the appropriate certificates to qualify for government financial subsidies.

Bharatbhai and the youth then established Ekta Yuvak Mandal. "We realized, as a group, we could be effective in improving our living conditions," Bharatbhai recalled. "We could seek support and guidance from Saath and as a peer group guide the youth of SMC." Ekta Bachat Mandal promoted a savings and credit cooperative, which disbursed loans for business purposes and personal emergencies. "People used to take loans at exorbitant interest rates of 10 – 20% per month and fell into ruinous debt traps. Ekta Bachat Mandal gives loans at 2% per month. This went a long way in getting affordable credit," he said.

---

[12] Described in Chapter 8, Page 96

Bharatbhai and other members of Ekta Yuval Mandal participated in the relief and rehabilitation of the riot victims in 2002. "We went to the camps to supply food rations," he recalls. "I was really afraid when our car, with a Durga (Hindu Goddess) sticker, passed a Muslim funeral procession comprising of a large angry crowd in Juhapura. The other passengers in the car were my colleagues in traditional Hindu attire. A small spark could have led to an attack. Luckily nothing happened and we passed safely," he said. "But it remains the scariest moment of my life. The experience of working with Muslim victims made him realize, "The big fish cause riots for vested reasons. The victims are innocent people caught in the crossfire."

Bharatbhai then got involved in the relocation of families affected by the Sabarmati Riverfront Project and the Metro Project in Ahmedabad. "I make sure that they get compensated as per the government guidelines," he said.

When asked to summarize his 28 years of work with Saath, Bharatbhai reflected, "The supportive environment, lack of discrimination, and participative decision-making encourages learning. I practice the same principles with people I work with. I have become a respected leader in my community and chawl." When asked about his future plans, he said, "I want to start a social enterprise to take care of the elderly."

Madhuben Parmar

Mature and wise, words that accurately describe Madhuben Parmar. 43 years old, she is the Chief Operating Officer at the Saath Savings and Credit Cooperative Society where she manages 26,000 customers, 11 branches, a loan portfolio of 17 Crores, a savings portfolio of 17.5 Crores and a team of 101 people. She has studied up to the 12th standard.

"My association with Saath began when I was in 8th grade and started attending the supplementary classes run by them in Guptanagar in 1992." Madhuben recalls, "When I was doing my 11th and 12th grades, I started teaching in the supplementary classes of the 1st to 7th grades to earn extra income." She got married when she was studying for her 10th grade. Her father passed away when she was in the 12th grade. Her mother, a construction worker, was unable to meet all the expenses of educating her two sons. Madhuben had to contribute towards the income. She had to give up her dream of further studies.

She officially joined the livelihood programme at Saath in 1996. "I was 22 years old and scared. My biggest challenge was to convince people, especially men, to take me seriously," she says. "I did that by delivering on the commitments I made at various meetings," she recalls. Residents of Guptanagar started trusting and respecting her. Subsequently, she became the secretary of the Sakhi Mahila Mandal, a women's CBO, whose main task was convincing residents of Guptanagar to join the Slum Networking Programme (SNP)[13]. "I went about it methodically." she recounts, "I identified local leaders and convinced them of the benefits of SNP by taking them to Sanjaynagar, where SNP had been implemented. I took them to meet the AMC officials and showed them films of similar work done in Indore."

"In Guptanagar," she explains, "There were seven different communities and the women did not talk to each other because of caste and cultural myths. We reduced these animosities by taking them regularly on outings where they had to interact with each other and eat their meals together. Gradually, the women started understanding that they had similar aspirations and challenges." These women bonded so well that they confronted their menfolk who were not allowing them to go for an overnight picnic. "We told the men that if our husbands and families did not object, why should the others interfere. It was a watershed moment for the women in Guptanagar."

When Saath decided to start a savings and credit group in Guptanagar, Madhuben took the lead. "We formed seven groups of about 200 women each, and appointed a woman leader for each group. During our meetings, women expressed the need to save and get loans at reasonable rates," she recalls. For the first three years, members of Sakhi Bachat Mandal just saved. They then became eligible for loans. Madhuben had to learn the intricacies of annual filings of Sakhi Bachat

---

[13] Described in Chapter 3, Page 26

Mandal with the district cooperative registrar. "I had to understand what was an official audit and, collate all the information required by the regulators."

She recalls how she learnt management and leadership by observation, reflection, and practice. "As a community worker, I understood the challenges of the residents. When I became manager, I designed my programmes to meet the requirements of the community. I understood the importance of planning, monitoring, and getting feedback. I had to motivate, my team, make them effective and sensitize them to the needs of participants. In the process, I learnt leadership skills. I understood the meaning of a long-term vision and the systematic steps required to reach a goal. I replicated the nurturing work culture of Saath in my programmes and, found that my teams performed better. This reinforced my belief in the values of transparency, respect, and accountability."

She has been involved with the financial inclusion programmes at Saath from the beginning. "I liked the cooperative spirit of savings and credit groups. I graduated from simple book-keeping to understanding balance sheets, share capital, and dividends. On the operations side, we made parameters for appraisals, affordable interest rates, loan amounts, and repayment schedules. I understood the value of computers, learnt the accounting software, Tally, and promoted the digitisation of accounts. I recruited and trained staff for field work."

Madhuben was instrumental in identifying the synergies of Ekta, Sakhi, and Sanklap savings groups and played a pivotal role in merging these groups into Saath Savings and Credit Cooperative Society.

Reminiscing her 25 years with Saath, Madhuben feels that she has made a difference. "When we started the savings group, there was a staff of 4, now there are about 100. From 130 members, we now have almost 25,000 members. I have made an impact in the lives of these people. I was offered a job by Fullerton India, with better pay, but chose to serve the communities. My husband and children are proud of my work. My mother-in-law talks appreciatively about my work in our village. I want to spend the remaining part of my working life in strengthening the Saath approach of working in partnership with communities."

Divyang Bhatnagar, CEO, Saath Savings and Credit Cooperative Society

Doing my post graduation diplomas and subsequently working in the UK left me feeling like I was in the wrong place doing something that didn't give me satisfaction. I needed a change.

When I returned from the UK I met up with a cousin who works for SEWA and mentioned to her that I was looking for a change. She suggested the social sector. I had no clue what the social sector was all about. She was working on a private assignment in Devgadh Baria, near Godhara in the Panchmahals and asked me to go along. This trip gave me an insight into the social sector; it was new and challenging – nothing cosmetic or artificial about it. I decided to look for an opening in the social sector.

I was told of openings in Saath. I applied for the post of an Accountant. Rajendrabhai asked me to meet him at his residence on a Sunday for the interview. He introduced me to the concept of microfinance. They wanted someone in their microfinance sector and were looking for someone who had not yet worked in the sector, who had no preconceived ideas about it, but could contribute fresh insights.

I was in a dilemma as to what I should do, as I didn't have a clue about microfinance, even though Rajendrabhai had explained it all to me. Based on a hunch and my instinct, I finally agreed. I like to have new challenges. It would be nice to be doing something good, not the run-of-the-mill type of work. I accepted it. At that time, Saath's microfinance section was undergoing a complete overhaul. So fortunately, I was part of the process.

The microfinance division started in 1994 was functioning in a typical NGO setup. There were two cooperatives, which needed to be merged. There were no professionals to look after the microfinance sector. I was directly offered the leadership position. Being in the overhauling phase, I was able to learn a lot and settle in. We had good consultants, Rajendrabhai, Chinmayiben and also people like Madhuben, and Parshottambhai; we made a very good team. Madhuben and Parshottambhai thought that they would learn from me, but it was the other way around. In the first year, I learnt from them the basics about the field – the kind of cultures and behavior of the communities.

I was in the UK during the riots; I had no clue about what actually happened. I learnt about the riots through the newspapers. Coming here and immediately working in the Juhapura area was a very, very good experience. My overall views changed.

When I joined there were consultants assisting us in the overhauling process and setting up new systems. Most importantly, Rajendrabhai and Chinmayiben gave me that space to explore and increase my current domain knowledge. The consultants too encouraged me to speak from my own experience, because ultimately it was going to help the organisation not in terms of the field or operations only, but overall development.

For the first year, it was all learning for me. Eighty percent of the time, I was the observer and twenty percent I was a direct contributor. From there on we worked in teams. When I say I faced challenges, it was because people from EDI and different financial institutions were working under me; they were more knowledgeable – as I said, I had no clue. I learnt from them.

I have been able to cultivate the culture, which Saath inculcated within the organisation. And slowly and steadily promoted people from within our organisation. I give credit to the team. The supported me. We have grown steadily over the past ten years.

I would definitely encourage other professionals to work in this sector. But I would tell them that if they wanted to come for the money, "don't come." I too had faced the dilemma regarding my finances, because I have a family to support. I would not say it is very little, but compared to the corporate sector, it is far less. What you will definitely get is a sense of satisfaction, and/or salvation.

I was a student in the UK and worked part time with corporates. But it was all about business, nothing to do with human interaction. Although there were philosophies during the start up, eventually it all fell apart. I feel that those who look for complete satisfaction, should, at least once in their lifetime, work in the social sector. Satisfaction level is really high if you choose the "right NGO." I have absolutely no regrets that I chose the NGO sector; that I chose Saath.

Shyam Singh, CEO, Saath Mahila Savings and Credit Cooperative Society

I came to Gujarat just after the earthquake in July of 2003 to assist with relief work and also hoping to land a job. I worked with a few agencies doing relief work.

In December of 2003 I landed a job with Abhiyan and my career in the social sector had begun. I am very grateful for my experience with them. I left them 2006 to join Pradhan in Delhi. A year after getting married I moved to Ahmedabad.

I got in touch with Chinmayiben from Saath, with whom I had been in touch with ever since I had attended the Leadership Training programme in 2004–5. I had a five-minute interview with Rajendrabhai, who told me that Saath was an action-based organisation. "I don't want to hear that due to a lengthy process no action has been taken." He had said. Saath needed someone to work on child's rights in the rural areas. An offshoot of the project was initially a savings programme. This grew into a women's micro finance savings and credit cooperative.

Rajendrabhai's mode of working was, trusting you to do your job. I barely met him for three years. I reported to Chinmayiben. Then when the project grew, I met him once more to discuss its functioning. His ideology was taking things step by step. When the leadership changed, I encountered problems with my style of working and was asked to outline a clear vision of the goal to be achieved. This was initially a huge bone of contention. We eventually managed to solve the issue by keeping some leeway and not being very stringent on 100% achievement of the goal set.

A while later, when the project came to an end, I was asked to work on the Human Rights project, as I was not able to draw my salary from the cooperative as yet. I agreed to work on the project, provided I could still run the savings cooperative, which was growing. I was able to manage the two.

Once again, there was a change in the leadership and my problems started all over again. With the new leadership it was once again, Saath's mission, project targets, and urban-based work. Merging the rural savings cooperative with the urban one was widely debated, and finally Rajendrabhai brought it to a halt. The rural cooperative would work independently.

Since 2017, I am working one hundred percent for the rural savings cooperative, which is able to support me.

If those that say there is no opportunity or career growth in Saath, I would like to say that if you want to do a job, then you work on a project; but if you initiate anything new, Saath definitely recognises it and supports it. To want to do something new you have to stay strong through its ups and downs.

I believe there are opportunities to explore and prove yourself in all organisations. Those who are unable to do so, very often end up finding fault and being judgmental. Those who are able to draw up a plan of their own and work with confidence, achieve it. Today, I have a sense of immense satisfaction knowing that I have created the Saath Rural Savings and Credit Cooperative. It is in a totally new area – rural women.

Zuber Mazudasud Sheikh

Zuber Mazbootmia Sheikh is a person with considerable gravitas and chooses his words carefully. He lives with his wife and three sons in Sankalitnagar, in the Juhapura area of Ahmedabad, which has, over the years, become the largest Muslim enclave in Ahmedabad.

Zuberbhai's 48 years have been eventful. Dire financial circumstances forced him to drop out of school in the seventh standard. He owns an automobile garage.

After the riots of 2002 in Gujarat, he worked in the large relief camp in Juhapura that housed families from twelve villages around Ahmedabad.

He was formally associated with Saath when we joined in the rehabilitation work at this camp. He says, with a twinkle in his eyes, that his association started much earlier, in 1984, when he met Fr.

Ramiro Erviti and Rajendra Joshi at various meetings, conducted by them through St. Xavier's Social Service Society in Sankalitnagar.

Zuberbhai became a prominent member of the Saath team involved in the rehabilitation of riot-affected areas in Ahmedabad[14]. He has been influential in building trust between the Muslim communities and us. In Sankalitnagar, he helped establish our preschool and financial inclusion programmes. He led the slum electrification initiative[15]. He was instrumental in convincing residents to go for legal electricity connections, as well as negotiating with the group, which supplied informal electricity, to counter their opposition. He established the Sankalp Mitra Mandal (SMM), a CBO in Sankalitnagar.

He was part of Saath's Samvad Community Video programme. Samvad made short community videos on issues, which affected the urban poor and screened these films to raise awareness. While working with Samvad, some of the films he was associated with were on the Government of India's Integrated Child Development Scheme (ICDS), which brought awareness about the multiple benefits of the scheme; a film titled Ekta, was made in 2004, when Diwali and Eid fell on the same day; it highlighted the similarities of the celebration of both festivals. He later worked in Surat where he filed a PIL along with a local right's group to prevent eviction of people living on government land without any compensation. Presently, he is associated with our work of rehabilitating people affected by the metro rail project in Ahmedabad.

Zuberbhai is passionate about Juhapura. During this interview, he got worked up when talking about the negative image conjured by the media after the 2002 riots, which described Juhapura as a place where weapons of all kinds were made and stored. He says that he went out of his way to show that the communities in Juhapura were like any other, when he interacted with officials from AEC, AUDA, and AMC.

---

[14] Described in Chapter 12, Page 147
[15] Described in Chapter 4, Page 46

A staunch Muslim, Zuberbhai is firm in his belief that his religion does not condone terror and the killing of innocent people. For him, true jihad is taking care of one's family, community, and being a good citizen. He said that a balance has to be found between personal laws and constitutional rights.

Towards the end of the interview, Zuberbhai explained that he has come a long way from being a seventh grade dropout. He is respected in his community and, is a member of the Shirat Committee in Juhapura, which promotes education. He is also a member of the Juhapura peace committee promoted by the local police. He said, "Now that the pressure of providing a livelihood for my family has eased, as my sons are earning members of the family, I can devote more time to the advancement of my community. I still have much more to do."

Ramilaben Shrimali

I live in Pravinnagar. I was a homemaker and had never worked anywhere. We had a water connection at our house. A woman who worked in Saath used to come and collect water. One day, she asked me if I was interested in working in Saath as a health worker.

On the 1st of July 1992 I joined Saath. In the beginning I was assigned to learn the ropes by observing how one of the health workers went about her duties. Subsequently, I underwent training in the various aspects of health, its causes, and remedies.

My first job was to conduct a survey of the area. Malnutrition was the single most outstanding issue. Both mother and child were malnourished. Beliefs and superstitions, along with poverty, were the chief causes. It also affected the immunisation of the children. We started educating women on nutrition, as well as distributed simple nutritious food to the undernourished women and children.

As the children were old enough to go to a playschool, we admitted them into the Aanganwadi. Here too, they were given nutritious snacks. I later ran an Aanganwadi centre; taught children.

Another disturbing health issue was tuberculosis. Most people shied away from disclosing their condition. We carried out house-to-house visits and checked on the prevailing symptoms. Sputum checks were carried out at the government hospital. We supervised medication and supplied nutritious food supplements. We assisted patients through the entire treatment – some were cured within six months and some within a year or more. The rate of tuberculosis has dropped significantly.

Over the years, I worked as a Community Health Worker in other areas – Juhapura, Santoshnagar, and Saraspur. I trained other women in the areas as health workers. We then formed small groups – "Matru Mandals" (Mothers Groups) and taught them women's health and childcare.

With water supply, good drainage, roads, and pucca houses replacing shanties, the environment has changed. Diseases and health issues have decreased considerably, education, and livelihoods have taken a turn for the better too.

Once the government sponsored RCH project started, I worked towards encouraging people to avail of government hospital services rather than private hospitals, which they were wont to do, even though the costs were high.

I underwent training on HIV/AIDS and have been part of the awareness programme.

Besides working on health issues, I participated in awareness programmes. Spread awareness about the various Saath skill training programmes as well as about the services provided by the Urban Resource Centre.

I worked with health related programmes for around eight years; HIV/AIDS awareness – one year, at the Aanganwadis – two years, Urban Resource Centre – five years, awareness for skill training – two years, and carried out surveys – health and education in different vicinities.

I am now working with the Home Managers project. My initial work in this project was to visit women who work as maids in multiple homes and explain the benefits of training and working as a Home Manager. I now have the responsibility of taking care of forty-six clients, solving their problems and providing a home manager to fit their job requirement.

Before joining Saath, I had lived within my four walls, hardly ever went out alone.

I have grown tremendously. Saath has seen me through my rough times and helped transform me. I am a confident woman; willing to learn and do new things, travel to any area of the city.

Bella Joshi – CEO, RWeaves

Though I was born in Africa, I was educated in India. Initially, I was in Mumbai and then moved to Ahmedabad where I met and married Rajendra.

Sometime in 2005, Rajendra asked me if I would like to work on a project on Child's Rights to Education. He explained that it was a joint venture between Saath and Jan Vikas and though I would be working for Saath, I would not be reporting to him.

Before we started work on the project, I underwent training in Coimbatore.

My job entailed working mainly with schools run by and for Muslims. The schools we covered were: FB School, Jamalpur, Anjuman Islam, at Astodia, Shama School at Dani Limda, and Diwan Balubhai School at Paldi. Other than introducing a different method of learning, we had to explain to the children their right to education and its immense value. We did this through various themes – painting, excursions, talks, etc. The project duration was for a period of one year.

In 2006, the Urmila Home Manager Project faced its first major problem. There had been a robbery in one of the client's homes and the Home Manager was suspended, with a case against her. I was approached to take up the post of Project Coordinator. The Sakhi Mahila Mandal (a Community Based Women's Organisation) had been running the project. I spent some time observing it's functioning and then joined the programme through the Sakhi Mahila Mandal. From my observations, I realized that we needed to make many changes for it to run smoothly. I discussed the issues with Chinmayi, to whom I directly reported, and made the changes after appropriate approvals. The training section of the project too underwent changes, as I worked closely with the faculty members and the trainees, as well as with inputs from clients.

As the project grew, another major change took place. The Sakhi Mahila Mandal was unable to handle finances, as well as the requirements of the fast growing project. In the meanwhile, The Livelihoods Services took off under Saath, and it was decided to shift the Home Manager programme under its umbrella. We introduced a helpline number for clients. We also introduced an application form for clients to fill in and sign.

Other innovations during that period included the Urmila newsletter. This was a compilation of snippets of other programmes in Saath, information on the new batch of trained Home Managers, as well as interesting recipes, especially during festival time. A Management Fee was introduced to cover general administrative costs.

We also made it mandatory for the Home Managers to be paid by cheque. They opened personal accounts and soon became familiar with banking procedures.

Representatives from the National Skill Development Council visited us and took a keen interest in the training and working of the programme. Representatives from the Tata Trust too visited us.

Presently I am working with Rweaves, marketing Tangaliya and Patola products. Dedadra village in Surendranagar district was known for its Tangaliya art of weaving, very beautiful and intricate. I had had the opportunity of visiting them before I started working in Saath. Because of the work done by Saath during the earthquake in the area they had discovered both the Tangaliya as well as the Patola weavers. It was a very dismal situation, as only two Tangaliya weaver families had continued to weave. All the other weavers had shut shop and gone on to work in other trades, as their market diminished with the influx of textile mills.

At Saath, we decided to try and help both the Tangaliya and Patola weavers to market their products. Initially, I brought material from them and sold it to staff members and friends. Then we held a small exhibition with whatever was available with them; it was a success, both products sold. People asked if they could give orders for patterns and colours of their choice for the Patola work. This started the personalized orders. From there we went on to design the marketing logo and name – Rweaves.

It has been a very enriching experience. We have managed to get more artisans weaving once more, and have expanded the product line. Today, we are selling the products internationally online.

I have seen many ups and downs, hardships and heartaches, through the years, working closely with those in need. Today, my life has meaning. I carry the feeling of immense satisfaction, knowing I do not live my life in vain.

I would like to let professionals from all walks of life know that the social sector is in need of their services. It is no longer a "charitable" service. It is all about development, working alongside those willing to work and in need of guidance and a helping hand. The feeling of achievement and satisfaction goes way beyond just building up material assets; in many ways it shapes us into caring and happy beings.

Paresh Sakariya, Senior Coordinator, Financial Inclusion Programme

I have been working for 14 years in the NGO sector. I was born and grew up in a village in Rajkot district. My parents are farmers. My grandfather, as well as my paternal uncle, had been a Sarpanch in our village. Watching them attend to the people, and even cattle, in the village, as well as in the surrounding villages, I felt that I wanted to do something similar.

It was after the earthquake. I worked for one year with Gram Swaraj Sangh based in Rapar. I then worked with Cohesion Foundation. I looked after the entire functioning of their newly acquired Navsari office. It was during my time here that I got married and wanted to move to a big city.

I first encountered members from Saath at a Life Skills training programme. The next time I visited Ahmedabad, I was introduced to Rajendrabhai, who interviewed me. I joined the Life Skills training programme at Saath. After three months, I was given the opportunity to start a youth mobilisation programme, for the training programmes. I moved on to liaison with government departments for the various training programmes offered by them through different schemes.

I liked working with the youth and was given charge of the Youth Force programme. We created youth groups and put them through skill training programmes. I am happy to say that most of the youth that were part of this programme are doing very well for themselves today.

I am associated with the Urban Resource Centre project. I am presently working for the housing project, which is spread over four states – Jaipur in Rajasthan, Ranchi in Jharkhand, Raipur in Chhattisgarh, and Banaras in Uttar Pradesh.

Other than these major projects, I was also briefly associated with the women and children's projects.

Today, I am a member in a number of committees, Strategic Management and Human Resources, to name a few. I have also represented Saath live on ETV.

Saath has given me opportunities in diverse spheres. I have learned a tremendous amount in Saath. I didn't know how to write proposals, reports, or even operate a computer. I am now very comfortable with all of it. The only weakness that I have not been able to overcome is the English language. I have to depend on others to do my English reporting and correspondence, as well as converse with funders. I am working at it.

My confidence has come from very good guides and teachers within Saath. Rajendrabhai, Chinmayiben, Keren, and Niraj, to name a few, have all been instrumental in my growth and confidence. The transparency and freedom to work has also played a large part in learning to deal with all situations.

My message to those asking: what have you done in your life? I cannot tell them how much money or wealth I have accumulated; but I am proud to enumerate the ways I the have helped, assisted, and guided those in need. The feeling of achievement through service is incomparable. Being remembered and acknowledged by those you have worked with, or for, is a reward in itself.

Kruti Jhaveri – Livelihoods Coordinator

After my graduation, I worked on the Solid Waste Management project at NID coordinating with AMC and Microsoft. I worked with student groups and then decided to move on.

I came across an advertisement from Saath and applied for it. The last three years have been a great learning curve for me. I joined as a Programme Head. I was not very sure if I could actually take up that position and deliver, mainly because I was expected to work with people who had a lot of experience, who were in the field for 10–15 years. However, Rajendra Sir and Nirajbhai were positive and encouraging – "you will be able to do it" is how they put it. That's how Saath happened. It wasn't planned at all. In fact, before I joined Saath there was very little I knew about the social development sector, because I came form a completely different sector. My background was urban management planning, a very different development sector.

The environment Saath provides is very conducive to growth. During my three years here I have definitely grown. Backgrounds or educational levels do not matter. The kind of support you get through people working here, many since inception – 29–30 years, is very encouraging. New young entrants too contribute with fresh perspectives.

I look after the Livelihood programmes. Earlier, we used to work a lot on non-traditional livelihood programmes and youth employability. But in the last few years, in collaboration with Universities, we have been able to develop new curricula. During the past three years, we have been able to develop curricula for 10-year olds and upto 40–45-year olds. We are now strong enough to take up the training of trainer's programmes. We work with young girls, 18–25 year olds, as well as with up to 40–45 year olds, men and women.

I come from a sort of privileged background, there are many things that I just take for granted. Saath has opened the doors to a different world where people go through a daily struggle over things and situations we take so casually. Joining Saath has definitely changed me a lot as a person. Because now I see what real struggle is, in spite of having a lot of capabilities, it is just the lack of opportunity that prevents personal growth. That is what we strive to deliver, day in and day out, through life skills training programmes and related activities we hold for our students.

The work we have been doing is good and solid. However, we have never worked at getting media coverage for most of our programmes. We do not associate ourselves with any one political party or group. We associate directly with the community. I definitely think, that for the next 10–20 years in order to grow, marketing is essential. There is so much social media where we need to project ourselves, put Saath in the limelight by letting the world know what we have accomplished in the market for 30 years and for the next 30 years what we plan to do. This is something I feel we can definitely strengthen.

I have taken up branding of all our centres across all our programmes. Working for a particular programme, one tends to work in that particular silo only. I have worked on streamlining and inter-linking of all the livelihood programmes. We are now working on uniformity of inter-linkages across all Saath's projects.

Step by step for each project, in year one, we take up audio visual content development, year two has theory development and year three is when we have a complete set of books where I can say Saath has its own curricula and this is something we follow.

Now three years down, we are at a stage if I want to tie up with any schools I can say that this is what we have developed and you can implement it, make it a part of your programme.

In 2016 and 2017 we worked to develop a Science, Technology, Engineering and Maths (STEM) curricula. It is basically activity-based learning approach for girls – school-going and school dropouts. It's an American concept, new to India. Only the international schools follow this curriculum. We worked with STEM experts, Jugar, from Baroda. We have developed curricula for 10–14 year olds, 14–18, and 18–25 year olds. We are now implementing it in 3 schools in Ahmedabad, Government-run and private, mainly for low-income groups.

We are in the process of developing a business model from the STEM curriculum. We are in collaboration with Duke University for summer interns. They help in designing workbooks, a facilitator's manual, and worksheets.

In our Women@Work programme, we work with female school dropouts and try and get them back into the formal education stream. For those in school who lack confidence to take up technical courses in higher education, this curriculum helps. It is not theoretical. It's just a set of 30–40 activities, which bring conceptual clarity in their day-to-day exercises and studies.

For most of our programmes we do a pre and post impact assessment through different activities for each programme. We have also been able to develop pre and post assessment guidelines for every programme. We are now able to track the social or educational impact with an overall understanding of day-to-day activities. Through data collection we are able to pinpoint small meaningful details. We have incorporated life skills in all our programmes.

To sum it up, there is uniformity and standardization in all our training programmes, for all training centres. This is how Saath's training centres can be identified. This has been an uphill task, as our trainers are mainly from the communities we work with.

Venugopal Agarwal

I am Venugopal Agrawal, an Architect and Environmental Planner by qualification. As a student, I studied buildings and urban design. After completing my education, I worked at MEGA (Ahmedabad Metro Project) Co Ltd. Working for a government company, I got to understand how a top down hierarchy works and implements decisions and designs.

After a year and a half with the metro project, I was offered a position as program manager at Saath Charitable Trust, to work on projects related to affordable housing. Although I understood basic concepts of housing from my architectural and planning studies, this was a completely new field and area of work... The most interesting part of the offer was that I would get a chance to work from a bottom up approach rather than a top down one. Housing also interests me as it is an important need that arises out of the growth of cities and urbanisation. I grabbed this opportunity with both hands.

As a program manager, I was responsible for the operational design and implementation of the Housing Resource Center Programme. This programme was being conceptualised when I started work. I was given the complete freedom to mould and shape it in the way that I saw fit. We undertook the task of associating and participating with 2 major government programmes – Pradhan Mantri Awas Yojana and ULC Validation. We focused on helping people access benefits under these programs. We conducted research, based on our experiences, as we tried to help people access these schemes. All our work was done with an exploratory and a trial and error approach. This allowed freedom to make mistakes, learn, and improve, which was a great thing to have. Not many jobs are capable of providing such a working environment, which is the greatest benefit of working at Saath.

Professionally, this freedom has allowed me to explore affordable housing in greater depth and detail than any other job could have. There is no proposal or idea that I, or someone else has had to which we have had to say, "No, let's not explore this." Constraints of time and resources have always applied, but that has been the only limiting factor, never anything else. I have explored ideas with scholars, real estate developers, people from the community, government officials, and my colleagues as well. While all of them may not have come to fruition, I have always had a clear line of communication through my association with an organisation that has confidence in the work that it has done, it's approach, and the results of its work.

As an employee, I am happy to have found an organisation that has similar values and understanding to those that I have always had. Several of my perceptions, values, and ideas have been transformed as I worked here. I have never had to compromise in any way. And this is the highest personal fulfillment that I could have asked for in a job. I think that being an NGO that is committed to better and integrated, holistic change in society has given Saath the ability to provide me that fulfillment. Working in an organisation that is not working for its own gain or a particular person's gain, gives one a great deal of personal and professional freedom and satisfaction. Finding an organisation where the focus is on the work and the quality of the work, rather than turning a profit, is one of the best things one can do while planning a career.

Keren Nazereth, Former Executive Director, Saath

A Professional, A leader – Always a Work in Progress

I studied Psychology in college and then went on to do my Masters in Social Work, from the Tata Institute of Social Sciences. When I joined Saath in November 2008, I was 3 years into what was the beginning of a career in documentation and content development in the development sector. A friend put me in touch with Rajendra Joshi, Founder of Saath, and he responded immediately. I was interviewed by Rajendrabhai and offered the job.

Saath to me was energy, dynamism, other young, driven people like me (if I may), who wanted to do the best they could. It was the professional working alongside the community in a symbiotic way, both equally essential to the process of change. Urban Slum Development was new to me. As a citizen, I had lived alongside a slum for the better part of my life and, to an extent, aware of the plight of an urban slum dweller. What I was learning at Saath was, how their lives could be greatly improved by working within the system and what it had to offer. I learnt how to translate government schemes into projects and to innovate; to not only change the system, but fight it, without being confrontational.

It did not matter that I had not done a course in urban slum development. What mattered was that I cared about the work we were doing and the communities we were working with.

I started out working in the Research Documentation and Communication (RDC) cell. A department that did not have much visibility, even within Saath. When you've been given a department, a designation, but no completely clear outline to work within, you create. I think that was the idea all along, to give people a space to make their own. I did, slowly. RDC focused on documentation for every project, writing reports, proposals, updating the website, getting Saath onto social media, online funding platforms, supporting projects with surveys, editing, set up a robust internship program. We grew from one person to two in a year and then a department of three.

I loved Saath and Saath loved me back. It was the energy, the support, the openness of the organization that had me giving it my all. There was no task too small, no challenge too big.

I was chosen to participate in a leadership program – the only young woman in a group of 12. This wonderful leadership journey eventually led to becoming a Co-Director, along with another colleague, Niraj Jani. In 2012, I went on to become Executive Director, Saath.

When I joined Saath, I had absolutely no notion the paths I trod would lead to becoming ED. This was one of the most fulfilling, yet toughest time of my career; my learning curve was the steepest. As ED, I had to almost magically be an all-in-one – approachable, a listener, a leader and a follower, a motivator, stand apart and yet be a part of the team, and most important, be accountable to and for the organization. It didn't come with a handbook. You learn on the job.

As an ED, the weightiest part of the job was being a constant liaison with different partners, developing new programs, innovating, supporting the program managers, balancing between the old whilst ushering in the new, accountable to the trustees, bolstering strong governance and accountability, keeping the humour alive, maintaining positivity when all seemed lost. My weakest points were delving into the books of finance and not letting my aggression get the better of me. I've actively worked on both these aspects, if it weren't for this experience, I'd have never known my drawbacks. I am still a work in progress.

Taking over from the Founder, had tremendous challenges. It was because of the mentorship of Rajendrabhai, Chinmayiben, Gaganbhai, the support of the community leaders like Devuben, Madhuben, Zuberbhai, Gopalbhai, Bharatbhai, Sanjidaben and the professionalism and friendship of many colleagues that I became ED. At 28, I experienced what very few professionals get an opportunity to experience – head a large, respected, impactful, established, non-profit in India. I made mistakes, but genuine forgiveness and support paved the way for all successes that followed.

In 2014, I informed the Board of my decision to move on. I wanted to follow my passion, working for animals. I had been offered a role in an organization that do just that. This was a tough decision, and it was not made lightly.

In hindsight, I have few regrets; learnings, many. As professionals, we tend to focus on the job, work, designation and its role. It is not often that a professional, no matter how well qualified, gets an opportunity to take up leadership within the organization that they work in. This may be the biggest flaw in the Indian NGO setup. Concerted efforts, like Saath's, to create second line leaders – next generation of leaders, go a long way in energising the organisation with an infusion of new ideas and fresh inputs.

My journey in Saath, a brief 6 years and 10 months was however packed with opportunities at every turn, and I took them. For me, every opportunity was one to learn, to take risks (with a lot of support), and to change – myself, my environment.

I will always feel a sense of ownership towards Saath. The professional and personal in me were nurtured with so much thought and care. I take the foundational values with me wherever I go – participation, leadership, integrity, openness, and most important, allowing everyone to make mistakes.

Niraj Jani, Former Executive Director, Saath

I grew up in Vadia village of Amreli district in Gujarat. My mother was a teacher and my father, a doctor.

After completing my graduation in architecture my first experience of working with a government project did not bring me satisfaction.

I then did my masters in Urban Planning, at the CEPT University, Ahmedabad.

While studying for my Planning degree, I explored the Gulbai Teka slums in my spare time. I gathered around 50–60 kids and taught them. Just meeting and interacting with them made a difference. I learnt that very small interventions could bring about change in a big way.

After graduation, I approached multiple organisations in Ahmedabad. Subsequently, I came to Saath and met Rajendrabhai. I spoke to him about my small interventions in Gulbai Tekra. He took me to Vasna and introduced me to the Sakhi group. Chinmayiben accompanied us. He introduced me to the work being done there and told me that if I was interested I could join. I decided to join.

I did not know about the general work at Saath, but I felt that this was an area in which I wanted to work. I felt that I could help with planning of projects. My entry into Saath was as a Project Coordinator for the Urban Resource Centres. This was in June-July 2008.

There were three Urban Resource Centres functioning within a structure created through experience. It was fascinating; I saw a strong community involvement. I started working with the team, Devuben, Yakubbahi, and Gopalbhai, the three Centre Managers. I was involved in restructuring the project with the funder. I enjoyed the work. We had proposed the idea of Urban Resource Centres to the Government of Gujarat. I worked closely with Rajendrabhai and learnt how to deal with the nitty-gritty of a working partnership with a government department. I became comfortable working in the team and learned how to deal with the government.

Another important learning point was dealing with community issues and working with them. I accompanied Chinmayiben on her visits to meet people, discuss issues and help bring about solutions.

Education was a field I was interested in. Being in the field of construction, I knew the situation of the children of construction labourers. I then worked towards an education programme for the children. I was able to convince developers and funders, and finally the Child Friendly Spaces programme for construction worker's children was born. The support system within Saath made it all possible. The difference between Saath and other organisations is not only the capacity of the individual, but the freedom and space everyone enjoys working in, as well as the constant nurturing of new inputs. Today, we reach out to more than a thousand children in this sector.

Another area I would like to highlight, is housing. As housing is my field, I worked with Rajendrabhai and developed the affordable housing programme. This entailed contacting and talking to various developers. I took the lead in designing the project. In Saath, we learn through our thought processes and experiences.

The transition from being part of the organisation to heading the organisation, was step by step, beautifully done; I didn't even realize it was happening. A Strategic Management Group had been formed, which consisted of project heads. Gaganbhai guided and trained the group. I was informed that I was part of it and agreed to attend the meetings. Subsequently, when Rajendrabhai decided that he wanted to step down, Keren and I were made co-directors and groomed for a period of one year under him. During this period, we were able to pinpoint our individual strengths and weaknesses and work on them. We took up responsibilities in the areas of our strengths. This prevented any kind of hierarchy.

Subsequently, Keren became Executive Director and I was the Co-Director. Since I was not responsible for the overall running, I spent time with the coordinators understanding the workings of each project in depth. I leant how things get done; it is not just by telling someone, designing or detailed planning. One has to get into the overall details and sort out the nitty-gritty issues, one by one. As I see it, everyone comes with so much energy and positivity, all geared up to work, and only need to be guided in the right direction. This experience is helping me tremendously today in terms of operations.

After Keren left, I took over as Executive Director. Rajendrabhai plays a supportive role, which is so important. The first challenge I faced, was when our office caught fire 5–6 months after my taking over. My immediate focus was to work on the rebuilding of the physical structure. There was tremendous teamwork inputs and support. I didn't get to feel solely responsible for the rebuilding of the office. Everyone pitched in physically, as well as with ideas and suggestions, which went a long way into getting our office into the shape it is in today. For me at that time, an architect and Executive Director, this experience showed me clearly how good the Saath team was, and that taking their design ideas, much against my architectural expertise, had been one of my best decisions. We were able to rebuild it with resources obtained solely for the purpose; staff members contributed a salary or a percentage of their salary… no project funds were diverted at all.

We started new projects and brought in more people. I started the process of recreating the Strategic Management Team, mentoring and ironing out details of projects. This dual role has its challenges, and I reach out to Chinmayiben and Rajendrabhai for help, as well as keep the Board of Trustees in the loop.

After ten years, the future of Saath, as I see it, comprises of three clear areas of stronger inputs and growth. The first is to build and strengthen the grassroots level, which has started with our method of planning and working today. The second is playing a larger role within the social sector, at a national level, policy level, and taking the lead in the national urban NGO space. Our role shouldn't only be carrying out projects, but sharing our knowledge. I see us doing that in a big way in the coming year. The third is, Saath's own expansion process. We have piloted programmes successfully some ten, twenty, and thirty years ago. Spreading them to different geographies and customizing them for the specific regions. We have done that in Jaipur, Ranchi, and Varanasi, where we now have our offices. We are liaising with a local organisation in Raipur and are in the process of handing over the project to them.

At Saath, we welcome interns from all over the world. I spend a very significant amount of time with each one, mentoring them.

Saath is an ecosystem – people working in the social sector on cooperative systems, enterprise models, working with the government, CSR projects, building a larger network of NGOs…..

Priti Shah, Former Coordinator, Research and Development Cell

I hold a bachelors' degree in Architecture (MSU, Vadodara) and masters in Urban and Regional Planning (CEPT, Ahmedabad). During the course of these seven years, I identified my inner calling, inclination and strength in social research and development issues rather than architectural design and physical planning – the conventional direction that my education would have led me to.

My journey with Saath started in the year 2000. While working as a Research Assistant in the Public Systems Group at the Indian Institute of Management, Ahmedabad (IIMA), one of my CEPT professors introduced me to Rajubhai from Saath. Little did I know then, that my first meeting with Rajubhai (as we fondly call Mr. Rajendra Joshi) would turn into a life long relation with him and the organisation. Working actively for five years with Saath from 2000 – 2005, my form of association with the organisation changed when I relocated to another city. I worked as a consultant on Saath projects on a regular basis, till I joined the corporate sector (Corporate Social Responsibility) in 2014.

Saath, in the year 2000, was looking for someone to document their journey thus far through a methodical and participatory approach. During the process of documenting Saath's journey, I got an in-depth understanding of the organisation's vision, program interventions and impact on the slum communities it worked with. The other factor that struck me was the openness, flexibility, ownership and fairness that seemed to be at the core of all its operations. Soon after, Saath offered me the job of setting up a Research and Documentation Unit (RDU). I was fortunate to work with a progressive leader who could envision this emerging need early on for any civil society organisation to grow and be seen more professionally.

The work RDU did was being a facilitator to the organisation's fieldwork while also establishing linkages with the external world. This, at the time, could mainly be divided into three categories: a) Setting up Management Information Systems (MIS) and documentation systems b) In-house process and impact studies c) External research consultancy.

I can best describe the role of RDU to that of a camera, which captures the field programmes and activities and puts it on paper for the external world, to fulfill the statutory requirements of periodic reports, presentations, proposals and impact. The other important role is to blow up these camera pictures to study the minutest details, in terms of monitoring progress, learning from experience, and making necessary changes in a consultative process. Internal evaluation and impact studies, process documentation, and vision papers were thus facilitated by the RDU. These also became important feeders in the organization's Planning, Monitoring and Evaluation (PME) exercises. Another enriching task was taking up consultancies for government institutions, academic and civil society organisations in areas of Saath expertise. This helped us immensely in broadening our vision on macro issues, which were then translated to grassroots implementation. Such linkages and partnerships were also a form of passive advocacy.

Post 2005, my understanding of the sector was further strengthened being a part of the conceptualisation framework of some innovative projects such as, the Urban Resource Centres, as well as impact and evaluation studies of some of Saath's key programmes.

Another important phase in my life was when I tried social entrepreneurship, and Saath, willingly gave me the necessary space and freedom to operate under its banner. Though the experiment was only partially successful, owing largely due to my constraint of being distantly located from the place of action, the learning helped in formulating a more successful and workable model in my future projects.

I am a firm believer in the symbiosis of personal and professional life. And my years at Saath have been an epitome of this balance we all strive for in life. A professional approach to work, and yet an empathetic view towards all employees and stakeholders helped us all stride through our professional or personal difficult phases in life. Saath practiced flexible work hours, time offs and working from home options, which enabled me, and several others like me, to fulfill personal and professional duties well.

While interacting with the slum women for work, I personally drew immense strength through their grit, determination and enthusiasm in the face of all adversities to fight my own challenges. Working with multiple stakeholders and a participatory approach was again a learning experience, in terms of capturing the strength of each institution and valuing positive partnerships for collective growth.

The most challenging and fulfilling part for me was to break down complex professional jargon into simpler, easy to understand communication for the slum women who coordinated the slum programmes. Making them learn the ' Logical Framework Analysis' was one such distinctly cherished achievement for me. Here, I would like to mention that when they learn, you learn more…mostly through unlearning.

A 'bottom-up approach' to project planning and implementation, in practice, made me understand the wealth of knowledge that was available and needed to be captured and utilised from all quarters. It is a humbling experience when the simple wisdom from grassroot workers and your work with the community is actually helping you grow and capitalise on your expertise professionally. The constant learning through regular 'inward looking' exercises and innovating to meet the continually changing external environment, made Saath a relevant and progressive organisation to work with. Another learning for me has been to see the strength in a 'non confrontational' approach to development that Saath follows, antithetical to the general perception of NGOs as activists. Today, I see this being the key to good inter-personal relations too.

Saath fulfilled the seeker in me. I believe, if the synergies are right, the organisation and the person both benefit mutually from each other. Reflecting back, I realize, this was the melting pot for diverse people from various backgrounds and with incredibly different ideas and approaches, united by an inexplicable like-mindedness of purpose. Thus, understandably, we all had arguments with our leader, with teammates – and this healthy exchange of ideas, opinions and differences was encouraged and allowed to thrive, which made the experience so enriching.

To the young professionals, who have a quest for being socially relevant or finding a higher purpose in life, my advice is to associate with NGOs/ civil society organisations. The key, however, lies in choosing the organisation that is right for you. If your core values and the organisational values match, you will be able to establish a long-term relationship. Otherwise, there could be disillusionment.

For the past few years, there have been a growing number of younger people questioning the dogma of a good education followed by a secure high paying job. If you are a person seeking a higher understanding of the self or identifying your purpose in life, there is a bright chance, you may get some of your answers here. There is nothing more fulfilling than making your passion your career. It always amazes me how absorbing the NGO sector is. Your knowledge in any field can be put to optimal use to the benefit of society, and in the process transcending you to a different level. With the emergence of Corporate Social Responsibility in a big way in India, the role of NGOs has also been redefined and partnerships are sought after for their expertise in community development.

Believe in yourself, listen to your inner voice and dare to take the path less travelled!!!

Simea Knip, Former Researcher at Saath

For a long time, the concept of affordable housing has fascinated me. In the Netherlands, social housing is part of the welfare system. Everyone with a low income is entitled to apply for an affordable house, provided by housing corporations that are controlled by the government. I wondered, how the distribution of social housing would look like in a country without a strong welfare system. So when the chance to do research in Ahmedabad practically fell into my lap, I didn't have to think twice!

I am an urban sociologist with an architectural engineering background. When I came to Ahmedabad in early 2010, I was just about to finish my master's degree in Sociology. I came to India early 2010 and started working with Saath from March. After an interview with the coordinator of the RDC, I started writing my master's thesis at Saath. I was also part of the RDC team and jumped right into the process of writing and designing the annual report. Working on the annual report was a very good way to get introduced to all of Saath's work and projects.

I come from a country where everything is planned, cultivated, calculated, and regulated. We don't like to leave anything to destiny. So, to put it mildly, the first weeks in Ahmedabad were quite of a shock. I felt like a helpless small child and even going out for simple groceries was a big challenge. My frame of reference had fallen to pieces. The RDC team however, made it possible for me to make a soft landing in Ahmedabad. With advice on clothing, traffic, street food, places to go, things to do, what to avoid, where to get what, basic Gujarati phrases, how to wear a sari, how to negotiate through traffic without getting killed, how to behave, and, not to forget, a lot of recipes! I went native in a few months time.

After the first orientation period, I finished my research proposal and started with fieldwork in two slums in Vasna. Saath helped me in find translators, to assist me with taking the interviews. Everyday I would go to the URC in Vasna (on my scooty, an adventure in itself) to meet Devuben, the URC's coordinator. After a chai and a chat, her daughter and the translator would take me to the families that were open to an interview. I met some really special people during that time. It's amazing to realize how you're able to bond without speaking Gujarati or Hindi. Hands and feet are great means of communication.

What I think makes Saath a really strong organization, is the diversity in cultural background and social class that represents Gujarati society. Saath works with the people of the communities, which Saath's programmes serve. Through Saath I became aware of the strength of diversity. It's important to share stories and connect with people of different backgrounds. It gets you out of your bubble and into real life.

On a professional level, I have learned a lot about (affordable) housing systems in India. I learnt about Integrated Slum development and the Slum Networking Programme, a unique and successful programme developed by Saath. I learnt about how people incrementally build and finance their homes in slum areas of Ahmedabad. I learnt about government schemes for housing and about forced eviction of slum dwellers. I learnt about how NGO's work together with the government and the private sector. I learnt about the importance of partnering up with other NGO's and companies. And I have learned so much more.

Working at Saath taught me to work with what's available and to achieve more with less. I think it gave me a better sense of sorting what really matters and more importantly, what does not. Forget about expensive management courses or hyped trainings: Jugad was invented here people!

Working with Saath was a unique experience. It gave me a very special insight into Indian culture and gave me the opportunity to become a member of the Saath family. If you are looking for a comfortable ' tell me what to do' kind of job or internship, then you could get disappointed. However, if you have in mind what you would like to contribute to the community, then you will meet a lot of enthusiasm and support to achieve your goals. I'm very grateful for all the time, space, and support the people of Saath gave me.

I returned to Amsterdam in April 2012, after a great journey of two years with Saath.

Jabin Khan, Former Coordinator, Rehabilitation Programme

My formal association with Saath has been a short one; just under a year – May 2002 to March 2003. I am an architect specialising in Urban and Rural Planning. I had been with the Heritage Department at the Ahmedabad Municipal Corporation, before joining Saath. My work was to conserve old buildings.

I joined Saath during a very difficult time. In February of 2002 our city had practically shut down due to the communal riots. I was unable to attend to my work of conserving buildings, primarily in the old part of the city. I was at home scared, wondering what I should do, when Saath contacted me. They wanted me to help in the rehabilitation of the riot affected families. I must add, my job interview was a short conversation with Rajendrabhai. No job title or salary was discussed. I was promised that my association with Saath would definitely change my attitude and outlook towards the underprivileged.

I count myself very lucky to have gotten this opportunity so early in my career; the opportunity to make a living doing work that enhanced my values and gave meaning to my life..... something too idealistic for a corporate setup. Working at Saath there was no sharp distinction between work and the rest of my life, between my interests and my job description.

I cannot say I had the skills for the job; it was a gut feeling that helped me take decisions. It was all with good intentions and so it was natural that often I walked away feeling that the world had become a better place, for some people in despair, because of something I did. That was tremendously powerful and motivating.

The people I have met in this line of work are among the most wonderful friends and colleagues anyone could have. Their influence has made me a humble and a caring person. Working at SAATH gave me more responsibility and authority very early in my career, something that would have come much later in the corporate world.

While the work with Saath had its own challenges, for me it was a wonderful opportunity. I had been very nervous, but I guess with all the support from the organisation I justified my role, and in return, got more and more responsibilities. Most of the tasks and problems I faced were extremely challenging, but I was often left with no choice but to attempt to address them, even though the skills and resources available were not sufficient. This nature of work was my biggest learning.

Saath has left an everlasting impact on my life. I am grateful to everyone at Saath who believed in me and trusted me.

Ushasi Sunandita

When Mr. Rajendra Joshi, Sir, to me, called up and asked me if I could write about my experiences at Saath, it was an overwhelming moment. So many beautiful and cherished memories just flashed across my mind, and, when I finally did sit down to write about my experiences, it was indeed an ordeal to arrange my thoughts and pen them in a coherent manner.

I have done my M.A. in Social Work from Tata Institute of Social Sciences with my specialization in Medical and Psychiatric social work. I completed my M.A. in 2002. In 2003, I shifted to Ahmedabad, a city that is extremely dear to my heart as it gave me invaluable friends, mentors, and my most favourite work place.

I met Sir at a workshop, felt the work Saath was doing and the Integrated Development model they had adopted to be innovative, challenging, and, at the same time, very comprehensive and creditable, and expressed my interest in being associated with them. We decided to meet at the Saath office the next day, and my journey with Saath started.

The office had an informal and warm feeling to it. And, Sir was so unlike a 'boss'-that's the first thing that in fact made me want to work with Saath. Conversations and discussions happened and I joined the Research and Documentation Cell under the exemplary mentorship of Ms. Preeti Shah, who gradually became one of my close friends, advisor, and guide. I, eventually, moved on to become the Coordinator of the Research and Documentation Cell.

I was a novice in the field of research and documentation. Saath gave me the opportunity and time to learn right from the scratch. Thus, started my learning. My mentor, guided, supported, and hand held me through every step, be it the basics of documentation, the nuances of writing a project proposal or effective, scientific, and unbiased research analysis. With time, I was given individual assignments that included project proposals, research analysis reports, power point presentations, donor reports, programme assessment reports, etc.

Saath is an extremely fun and warm place to work at, but, trust me, they are extremely professional and dedicated to their work. The field trips to the areas they were working with, meeting people at the field and conversation with them were so enriching and full of insights. It was an absolute pleasure working with the field team. I made some everlasting bonds with a few of them. The field team is very knowledgeable and has a deep understanding of the field realities. Their ability and courage to remain calm even in the most adverse situations, such as the Gujarat riots, and continue working for their mission and belief, is in fact a very important life lesson for anyone. The work done by the team during the riots is an example of their compassion, dedication, and professionalism.

It was indeed a kind of an adrenalin drive – meeting time lines, making project proposals over tight deadlines, keeping pace with ingenuous ideas.

While Saath is extremely professional in its approach to its work, it was at the same time home away from home from me, on the personal front. The working atmosphere here was very warm and congenial. There was no hierarchy that ever stifled my thoughts or voice. I was always heard patiently and given the freedom to work my way and put across my perspective without being misjudged – a very rare privilege in working environments really!

Saath gave me the freedom to learn, explore, make mistakes, correct them, make some more mistakes and learn from them – it gave me the pleasure of being myself without any fear or apprehension. Saath was and has always been family – I felt protected, secure and content there. At the same time, Saath gave me innumerable professional challenges to satisfy my quest to learn more, gain experiences, and do something purposeful with the people with whom I was working.

As a woman, I appreciate even more the gender policies of Saath. The flexible timings, the safe environment, the equal opportunities that it gave to each person associated with it, made the experience of working there so much more enriching and fulfilling.

I was professionally associated with Saath till 2009 (that's when I moved to Delhi), but on a personal front, Saath is always family-it's a never ending association, and I am so proud and happy to be a part of the Saath family.

Saath also breaks the myth that NGOs do not provide a challenging work environment or are not professional in their work policies. In my opinion, times have changed and NGOs give one the perfect opportunity to balance their professional expertise, home, and the satisfaction to be able to do something for the society. It is no more a 'jhola wali behenji clad in a salwar kurta' mindset.

Saath has always been close to my heart and it is indeed a pleasure and privilege to have been associated with them.

Chapter 17

My Personal Journey

In the early 1900s my grandparents migrated to Tanzania, East Africa. My parents and siblings were born there. In 1977, my parents decided to return and settle down in Ahmedabad. I was nineteen years old. Growing up, for me, India was a distant mythical country of villages with archaic customs and distant relatives. The only connection was through the recollections of my grandparents and, watching Hindi movies, which were quite popular in East Africa.

I was fortunate to grow up during a period when African countries were gaining independence, either through negotiation or armed revolution. While in school, we were closely exposed to the liberation struggles in South Africa, Angola, Mozambique and Rhodesia. The struggles of the stalwarts of these movements, Julius Nyerere, Nelson Mandela, Patrice Lumumba and Samora Machel were a part of our lives and a source of inspiration.

Being part of a comparatively privileged Asian family, I pondered on the inequities between the powerful white people, the industrious, but often exploitative Asians, and the oppressed black people. I also experienced the churning of political, economic, and social equations when Africans gradually took over the reins of power with its positive and negative effects. My friends from varied economic backgrounds were Asians, Arabs, Africans, and Europeans practicing Islam, Christianity, and Hinduism. This diversity, taken for granted, helped me evolve a secular world-view, which was encouraged by my parents.

With just my grandparents reminiscences and the Hindi movies I had watched, I was in for a huge culture shock when we returned to India. The vibrant democracy and independent media (The Janata party had just won the elections and Indira Gandhi had been dislodged) were like a breath of fresh air after the stifling one-party rule in Tanzania. I was amazed by the diversity of higher education options and livelihood opportunities. I gradually saw and experienced, first-hand, the inequity and cruel realities of caste, class, and religion. I graduated with a bachelor's degree in zoology in 1982. I worked for two years as a medical salesman and an export-import clerk. These jobs did not give me any satisfaction, enjoyment, or sense of personal fulfillment.

Looking for a change, I joined St Xavier's Social Society (SXSSS) in 1984. A voluntary organisation working in the slums of Ahmedabad. This was the beginning of my journey in the not-for-profit sector. The director of SXSSS, Fr. Ramiro Erviti, a Jesuit from the Basque region of Spain, helped me realise my vast potential. (He did this with many other students and colleagues). Fr. Erviti's empathy quotient was very high. He had a special brand of modus operandi. He would throw professional challenges at those he felt could deal with them. I very often thought I could not handle those thrown at me. But his belief in me, and his guidance, and empathy would force me take up those challenges and, in doing so, I discovered my capabilities. This process of uncovering one's potential is life enhancing. I have replicated it with many colleagues at Saath. With Fr. Erviti and SXSSS, I learnt the basics of engaging with youth, project management, and teamwork. The close interaction with vulnerable communities living in low-income neighbourhoods was the beginning of a life-long work-romance and commitment.

Unfortunately, Fr. Erviti passed way in 1986. I began to feel the new leadership and working style changes in SXSSS were regressive. I decided to move on and worked with two NGOs, whose focus was not direct community interaction. I realized that to fulfill my desire of working directly with vulnerable communities in Ahmedabad, and making a difference, I would have to venture out on my own. In February 1989, with my colleagues and friends, Robert David, and Pradeep Seth, I registered Saath, as a voluntary organisation.

Our vision and mission was creating inclusive cities through a holistic approach. We started engaging with communities and building relationships with people, which still continue. We built our team with members from within the communities and professionals who wanted to make a difference. We developed planning and review mechanisms, which promoted reflection and learning, and evolved a working environment, which brought out the best in people and led to pioneering programmes. Saath became a place where 'ordinary' people brought forth extra-ordinary results through teamwork.

Through our integrated approach, we learnt about pre-school education, preventive health, financial inclusion, livelihoods, and basic infrastructure and services. We learnt how to respond to disasters and displacement. We built partnerships with donor and philanthropy agencies, government, private sector companies, academic institutions and civil society institutions. We have understood the important role of the informal sector in making our cities work and the vital role gender plays in development. Most importantly, we have created an institution, which is making a difference in so many lives.

I am often asked why I chose to work in the not-for-profit sector. My answer has always been that this sector has allowed me to be creative and learn leadership, management and organisational skills, and to use these capabilities to make a difference in society.

The Not-For-Profit sector is diverse and offers many opportunities for personal and career growth. It encompasses health, education, livelihoods, financial inclusion, housing, construction, and agriculture sectors. There is a need for technical, managerial, marketing, research, and social development skills. There is a scope for both, generalists and specialists. The most attractive part is the personal transformation of people who live with dignity in very trying circumstances.

As a person, I have gained so much during my thirty-five years in this sector. Perhaps, my biggest gain has been understanding the value of empathy and how the practice of this value can have far reaching positive outcomes. At a people's level, I have learnt to be humble, patient, understand the other person's point of view, be non-judgmental and communicate effectively. At an organisational level, I have learnt how to motivate team members, establish inclusive management style, planning, reviewing and monitoring, gathering financial resources and using them wisely and effectively. I have learnt to engage constructively with people from all strata of society as well as national, state, and local governments, academic institutions, corporates and philanthropists. I am sure that I would not have gotten such a varied exposure, had I been working in the private sector.

Although, the not-for-profit sector offers lower salaries than the corporate or government sector, it is adequate. My take is that the rounded and considerable personal growth I have achieved, as well as making a difference in the lives of so many people, more than compensates for the lower financial rewards. If I were to be reincarnated, I would choose the same field of work without hesitation.

To young people who want to be change makers for a more equitable and sustainable society, my advice is that the development sector is a choice worth considering. You will get an opportunity to use and develop your creativity with like-minded people. You will discover your vast potential by contributing to making this world a better place to live in. The not-for-profit sector is growing in India, with the advent of mandatory CSR for corporates and, increasing philanthropy.

The seventh full edition of Bain's India Philanthropy Report states that, 'The philanthropic ecosystem has been thriving due to the combined efforts of both the public and private sectors. Overall, total funds for the development sector have grown at a healthy rate of approximately 9 percent over the past five years, increasing from approximately INR 150,000 crore to approximately INR 220,000 crore. While the government remains the largest contributor (INR 150,000 crore in 2016), its share in total funding has been declining steadily. Private contributions primarily accounted for the INR 70,000 crore five-year growth. Private donations made up 32 percent of total contributions to the development sector in 2016, up from a mere 15% percent in 2011.'

The current economic models are not creating wealth equitably. The rich are getting wealthier and the rest are getting poorer. According to Inequality.org, which has been tracking inequality-related news and views for nearly two decades, in 2017, more than 70 percent of the world's adults own under $10,000 in wealth. This 70.1 percent of the world holds only 3 percent of global wealth. The world's wealthiest individuals, those owning over $100,000 in assets, total only 8.6 percent of the global population, but own 85.6 percent of global wealth. I believe that this disproportionate distribution can be attributed to the importance given to individual financial gain rather than the general well-being of society and the environment. We require different models for creating and distributing wealth so that those who contribute towards creating wealth are better rewarded.

A collaborative model between the not-for-profit sector, the government, and the private sector can be a solution for ensuring a fair way of creating and distributing wealth by empowering disadvantaged communities, furthering corporate social responsibility, and by influencing and facilitating government programmes. We have shown how Saath has done this.

Notes from Experts

Mr Gagan Sethi, Chairperson, Saath Board of Trustees

Saath's journey over the last 30 years has demonstrated that professionals who have no philanthropic background can initiate powerful institutions. What has been important is to follow basic values and tenets of institution building. These are: build an organization with community engagement and empowerment as central to the DNA and build capacities of professionals and para professionals in sync with the growth of the organisation.

Having been witness, accompanier, consultant, mentor, board member and now, Chair of Saath, I give the credit to Rajendra who committed Saath to become a public institution rather than a personal, private, or family enterprise, which many non-profits have been reduced to in the country.

A commitment to such a process entails understanding, believing, and practicing the basic tenets of non-profit governance. Some of which are:

a. Collective transparent decision making, as a practice within the organization.
b. Demonstrate Accountability to its stakeholders, as a public body.
c. Consistent and constant investment in building second line leadership
d. Build an ecosystem of interdependent organisations to proactively meet new challenges.

Saath's growth from a small friendship board, to an accountable board, to a professional strategic board, is a story many founders could learn from!! As a founder, trusting and giving up control is not easy, especially when you have nurtured an organization from scratch. It requires a deep understanding that life of an institution is longer and larger than that of an individual. Thus, submitting oneself to collective wisdom and setting in place practices to see that the Board take ownership of the institution and, that it is a ROLE they play for the limited time they hold office.

The Saath Board has put in place a finance and governance committee, brought in a system of internal audit, to ensure both compliances, as well as transparency in operation. It nudged and nurtured the new CEO's to meet stringent and ambitious targets while ensuring high standards of reporting.

As Saath began to incubate other entities it had to sift through myriad multiple governance and ensure that it avoided conflict of interest. New mechanisms, therefore, based on principles of meta-governance are now being put in place.

The tenuous relationship between boards and executives is always a grey area. The Saath board and executive are constantly negotiating boundaries. This is well balanced by Rajendra, who holds twin roles.

The other area of Organisational Development, which Saath has been proactive in, is investing in Leadership building from within. The second and third layer in the Saath Ecosystem is, constantly giving opportunities to both professional staff and staff inducted from communities to go through trainings and reflection sessions. To push the staff members to take new responsibilities, there has been constant restructuring. An appraisal system and a gender sensitive policy encourage women to take leadership positions.

The board itself has set up a practice of reflecting on changing economic and social environment and, urges the executive to look at engagement with the corporate sector, government, and also respond to crisis situations.

It is through these processes that Saath commits itself to its Vision and Mission and, practices its values.

Prof Chetan Vaidya, former Director of the National Institute of Urban Affairs and School of Planning and Architecture, Delhi, Trustee of Saath

As per Census of India, in the year 2011, about 17 percent of the urban households in the country were dwelling in slums constituting staggering 13.75 million or 65.49 million population. It would seem the majority of these houses are inadequate and poorly serviced as revealed by the recent urban housing shortage figure estimated to be 12 million.

Housing for the poor has attracted the attention of the Central and State Governments since independence. The role of public sector for providing housing was most important beginning with the First Five-Year Plan. Slum clearance and built tenements were the major mode of housing provision for the poor. By the seventies, it was realized that the strategy was not giving expected benefits and during the 1970s, slum clearance was replaced by radical reforms recognising the contributions made by poor households in self provision of housing. This recognition led to provisions of housing finance and improvement of slum environment by developing basic and social infrastructure rather than actual construction of tenements. The 'Site and Services Schemes' and 'Self-help Housing' projects provided tenure rights to the urban poor facilitating their own contribution towards building a house at their own pace in an incremental manner. During the eighites, Urban Basic Services Programme was laucnhed. The programme had community participation and convergence as key elements. One of the shifts in the approach to housing and basic services was involvement of Community-Based Organizations (CBOs) and Non-Governmental Organizations (NGOs) in housing and service provision that continued to increase over the years.

By 1990s with larger economic reforms, perception of housing shifted from being seen as a welfare measure to contributor to economic growth and as an investment good. Environment for private sector investments in housing was facilited through relaxed legal, financial and tax regulations. The government's role was refined as a 'facilitator' from that of a 'provider'. A revised National Housing Policy was formulated in 1998 focussing on improving access to serviced urban land, housing finance and innovative technologies for affordable housing.

Beginning in the 2000s, especially the 11th Five Year Plan, private-public models to provide housing were promoted. National Urban Housing and Habitat Policy 2007 (NUHHP-2007) promoted various types of public-private partnerships in achieving the goal of 'Affordable Housing for All'. Housing provision was seen as part of citywide reforms. Funding for housing and infrastructure from the Centre was linked to city performance through mandatory reforms. The Government of India launched reform-oriented programmes such as Basic Services for the Urban Poor (BSUP) as part of the Jawaharlal Nehru National Urban Renewal Mission (JNNURM). Later, the Rajiv Awas Yojana (RAY) was conceived having 'Slum Free India' as its main focus. In RAY, NGOs were envisaged as a bridge between the urban local bodies and the people.

Since 2014, India's response to urbanization recognizes the international benchmarks as laid down in the Sustainable Development Goals (SDGs), and its guiding principle of 'Leaving No One Behind', the Paris Agreement on climate change and commitment to the New Urban Agenda (NUA). With increasing importance of the urban sector, over the past few years, India has stimulated the urban sector through launch of new missions covering Swachch Bharat Abhiyan (Clean India), Housing for All, Smart Cities Mission, Heritage Rejuvenation (HRIDAY). The objectives of the missions are to improve the quality of life in urban areas.

In pursuance of Government's vision of facilitating housing to all by 2022, the Pradhan Mantri Awas Yojana (Urban) – PMAY (U), Housing for All Mission was launched on 25th June 2015. It addresses urban housing shortage among the EWS and LIG category including the Slum Dwellers by ensuring a pucca house to all eligible urban households by the year 2022.

However, in spite of the its commitments, the current pace of urbanization alongwith limited supply of developed land and poor management becomes a cause for increasing existing city problems, including access to safe water and sanitation, and adequate housing. The urban poor comprise different groups with diverse needs and levels and types of vulnerability. Urban India is too large and diverse to have a common approach in addressing urban poor issues.

In this context, in the last thirty years, SAATH has taken up number of approaches of working with poor. The chapter on Basic Services and Housing describes SAATH's work on facilitating sanitation in a chawl, provision of basic services through the Slum Networking Project (SNP), improving access to formal affordable housing and the facilitating housing for migrants. Each of these experiments have lessons for urban India. SNP in two slums of Ahmedabad city involved improvement of physical infrastructure and community development. It was a path-breaking initiative. For the first time, slum development was undertaken through forming of equal partnership of slum residents, municipal corporation, industry and NGO. This was an innovative idea but sustainence of partnerships is a tough challenge. The partnership approach initiated by SAATH was a very innovative experiment that needs to be explored in different contexts including in-situ rehabilitation approach currently being promoted across country.

SAATH has rightly advocated incremental formal housing and treating it as housing continuum. It is also involved with poor communities to improve access of affordable housing. SAATH promoted "Griha Pravesh" as a social enterprise to create awareness about affordable housing. This initiative guided a large number of households to own affordable housing. Housing Resource Centres have helped migrant workers to access basic urban services and shelter.

In this context, work of SAATH for last thirty years with urban poor raises following issues:

a. How can the approach of providing community-led basic services to urban poor with limited land tenure could be replicated all over urban India?
b. Can incremental housing improvement be made a part of national housing program?

c. How can role of public housing agencies be changed from providing new housing to supporting owner-built housing in terms of guidance and technical assistance?

d. How can the housing finance companies be encouraged to provide access to affordable housing finance to urban poor?

These are not new issues for basic services and housing sector in urban India. However, these will require change in focus of national urban housing programs. Urban India needs a large number of initiatives like those taken by SAATH to solve its large and diversified problems.

Mr Dilip Chenoy, Secretary General, Federation of Indian Chamber of Commerce and Industries and former Managing Director and CEO of National Skills Development Corporation.

India is going through an interesting phase in its economic development. Reforms ushered in different areas since 1991 has made significant changes in the lives of people by increasing the participation of the private sector in the economy. The better GDP growth, especially in the last 15 years, has improved the living standards of the people in this country in a big way. Poverty has been continuously in the decline, but the huge challenge of an inclusive and equitable growth has also emerged. Government policies, including MGNREGA, and the ones implemented in the last few years — PM JAN DHAN Yojana, Smart City and Aayushman Bharat — have transformed the capabilities of the poor and people in the unorganised sector in handling the uncertainties of life by making them financially capable.

These will help in curbing the inequalities in the society, but there has to be a framework with this to assimilate all this with the challenges posed by rapid urbanisation so that the new India also doesn't have cities with areas divided on the basis of income levels. This is not an easy task and more than the government, this will require the participation of people across the board.

Saath's journey of three decades is an example of how serious work towards bringing various stakeholders together to power city development utilising the strength of the unorganised sector. The work has ensured fulfilment of the twin objectives of improving the living standards of people living in low income neighbourhoods while they are empowered to make their contribution in making the cities smart.

This book not only covers the good work done by Saath, but also provides food for thought in terms of what needs to be done to scale up this exercise and make it more relevant and a powerful mode to make our cities a much better place to live than what they are today.

While improving city life is important, more important here is to do this in an inclusive manner that leads to equitable development. This can happen only if the unorganised sector workforce and their families, which includes every segment that impacts people's life in a small or big way — rickshawallas to drivers, vegetable and fruit vendors to dabbawallas, just to mention a few amongst those playing a critical role in running the city life uninterrupted — have optimum possible health, education, housing and insurance facilities.

The role of social agencies like Saath is extremely significant here. Saath has already proven this by making an impact on the lives of over one million individuals by working in these critical areas such as health, education, livelihoods, housing and shelter, financial inclusion and disaster rehabilitation. This has empowered and enhanced the quality of life of individuals from vulnerable and low-income neighbourhoods living in our cities.

What Saath has done over the years in its own small way, needs to be turned into a big movement, towards structured initiatives with the help of innovative programs, in a major way, and this is not difficult task if all the stakeholders are brought together through public initiatives.

This book will certainly help in inspiring people to join the NGOs in their own ways and help in creating an atmosphere for social change that will lead to a more inclusive growth.

The real benefits of the social sector schemes launched by the government, including Smart Cities Project, will be realised only when they are able to drive economic growth and also simultaneously improve the quality of people's life through participation along with the utilisation of technology.

Saath has shown how the partnerships with communities, civil society, academicians, government, and also the private sector can come together to transform the lives of poor and vulnerable.

Skill development has to be at the core of all this as without raising the employability levels through training and development, equity and inclusiveness can't be taken to the desired levels. This is both a challenge and an opportunity for Saath and other organisations working in this area to take their work also to the next level.

While many organisations are active in this area, few have been able to scale to the extent that Saath has been able to. This ability to scale and to continue to grow if replicated by organisations in this space can truly lead to a transformation of India. I am happy to note that this book is set to make an effective contribution to this process.

Vijayalakshmi Das, Managing Director of Ananya Finance for Inclusive Growth Pvt. Ltd. and Ananya Vijayalakshmi Finance. Ms. Das served as the Chief Executive of Friends of Worlds Women Banking

To drive an inclusive growth of a country, financial inclusion is a key enabler. In India, it has been quoted as one of the key priorities of the Government since post independence days. Starting from nationalization of Banks to e-KYC, the main focus was financial inclusion. In spite of the impressive growth of the economy, post liberalization, we continue to account for one third of worlds poor. There is a huge gap between what our policies say and what is followed in the field. The latest effort by the Government is the Pradhan Mantri Jan-Dhan Yojana (PMJDY) under the National Mission for Financial Inclusion, announced in 2014, through which affordable access to financial services to the vast sections of disadvantaged and low-income households is to be ensured. The financial services went beyond savings and credit to insurance, payments, and remittance facilities. The success of the scheme in closing the divide between the excluded and the included is yet to be assessed.

Some of the major challenges even the Government is facing are physical and internet connectivity, higher percentage of dormant accounts, lack of officially valid documents to open bank account, high rate of illiteracy in some pockets, high attrition of business correspondents, etc. Unless they are addressed, financial inclusion will be a distant dream. Like Saath, many NGOs in India realized in the early nineties that access to financial services is one of the critical components of poverty reduction, both, in urban and rural areas. To mitigate this major constraint of access to affordable financial services, formation of Self Help Groups were the first step. Those keen on creating a community owned structure (only available options were co-operatives, federations of savings and credit groups as Societies) needed to experiment a lot and met with different challenges. These NGOs while recognizing the importance of access to financial services also gave importance to other areas – livelihood promotion, health, education, legal entitlements, policy interventions, etc., necessary for poverty reduction.

The chapter on Financial Inclusion narrates the struggle in setting up the **Saath Credit and Savings Co-operative Ltd** in 2010. The journey started in 1990 with the effort to set up a revolving loan facility for a sanitation programme, forming SHGs to address the need for working capital, medical and other emergencies, forming 2 Savings and credit cooperatives in 2007 and finally in 2010 the single entity Saath Credit and Savings Co-operative. **The two-decade journey of Saath is very interesting to read and spells out clearly the various hurdles one has to encounter while organizing the poor, especially around financial services. The success of this particular effort of Saath is dependent on the continuous growth of Saath Credit and Savings co-operative.**

Prof Amita Bhide, Dean, School of Habitat Studies, Tata Institute of Social Sciences

Community development is an idea that developed in the 1940s through several experiments in various parts of the world. This included work in developed countries but also includes experiments such as Sri Niketan led by Rabindranath Tagore in rural areas of Bengal, trying to combine social transformations with efforts to marry tradition and new science for scripting a vision for modern India. When the country became independent, community development was the strategy of choice for change in villages ushered by gram sevaks who would bring knowledge produced in universities to the villagers and supported by extension departments.

In the 1960s, there were efforts to professionalise community development as a method of social work. Prime among these is the definition of Murry Ross who famously described community development as the process by which 'a community identifies its needs/problems, prioritises the needs, identifies resources to resolve the problems, undertakes action vis-vis these needs and in so doing develops the collaborative and other capacities (Ross, 1967). The Gulbenkian Foundation played a critical role in formalising and privileging this approach by preparing a report that brought together prevalent knowledge about the practice of community development (Gulbenkian Foundation, 1968). Community development was thus a part of the promise of the new world post the second world war where it was seen as a bridge between the state and people; between scientific, modern knowledge and technology and people embedded in tradition; and between various groups of people divided by caste, class and privilege. The bridge was to be constructed by the efforts of professional social workers.

Yet by the 1970s, community development was given up as an approach. The shift away from community development was not just seen in the shift of development policies but also in the professional social work realm where the consensual approach of community development gradually gave way to the more political and activism-oriented community organization. Community development was critiqued for assuming that structural conflicts and power relations in village and other communities could be transformed into consensus and that benefits of development could be shared equitably by all members of the community. In most contemporary literature, community development is regarded as a thing of the past.

Ms Indu Capoor, Founder Director, CHETNA

For three decades, SAATH tirelessly worked on "real concerns" of poor and disadvantaged urban communities. When it started its work on the large Slum Networking Project (SNP) in Behrampura as early as 1995, they could have limited their work to the task given – building roads, ensuring electricity and sanitation (as there were no gutter lines in the slums). However, they observed that due to overcrowding, poor ventilation and unhygienic sanitary conditions; the health of the community remained poor. This is when SAATH stepped out of their comfort zone and started the "Community Health Programme." They were pioneers in piloting a strategy of Direct Observation Treatment for TB. WHO adopted the strategy in 1995 and entitled it as the DOTS; an initiative, which has gained immense popularity worldwide.

In 2001, SAATH actively contributed to the earthquake rehabilitation programme at Bhachau, Kutch supporting the government health team in providing the much-needed health care services.

Rehabilitation of the 2002 riot victims is another area of their tireless efforts in health management. Not only physically, but also emotionally and psychologically, they worked with traumatised women and children.

SAATH's work on HIV/AIDS awareness is also commendable. Their untiring efforts over the years are very visible in the transformation of the slums of Vasna and Behrampura.

The key feature of the SAATH health programme is the Community Health Programme implemented through an "Integrated approach" linked to livelihoods and savings programmes, which is now recognized as the most effective approach for improving the health of marginalized and disadvantaged communities.

Since 2014, the Government of India (GOI) launched the National Urban Health Mission (NHM), which has now merged with the National Rural Health Mission (NRHM) entitled National Health Mission.

NUHM envisages to meet the health care needs of urban population with focus on urban poor, by making available essential primary health care services to them and reducing their out of pocket expenses for treatment. This is envisaged through strengthening the existing health care services delivery system, targeting people living in slums and converging with various schemes related to wider determinants of health, like drinking water, sanitation, school education, etc. implemented by ministries of urban development, housing and urban poverty alleviation, Human Resource Development and Women & Child Development.

This is a great opportunity for SAATH to collaborate with the Government to expand their community health outreach. There are already urban health centers, free medicines and the scheme also has a provision for monitoring mechanism through Mahila Arogya Samities (MAS), which includes Urban Social Health Activist (USHA), Aganwadi Worker (AWW) and other women belonging to the community.

Based on the expertise and experiences of SAATH, this could be undertaken by them. Through extensive follow up and mentoring support, these samities can be activated. It has been CHETNA's experience that the participating women can be empowered and mentored to demand their entitlements, which have now become a reality, through NUHM.

Education Initiatives

SAATH initiated the implementation of their pre-school programme through Balghars based on the need and analysis of the situation in the Juhapura slums. The approach adopted by SAATH of active involvement and engagement of parents and teachers in the process of educating the children, was unique. As the Integrated Child Development Scheme (ICDS) is presently being re-formulated, there is an opportunity for SAATH to share the lessons learnt from Balghars with the Government policy makers at the state and National level.

The Child Rights for Change, a noteworthy programme that SAATH undertook had a very positive impact in the lives of children in the cotton growing Talukas of Viramgam and Ahmedabad. The Child

Protection Committees (CPC) become an institutional mechanism for addressing local child labor issues.

Through the Child Rights project, SAATH enabled the below poverty line women to form Self Help Groups (SHGs). Apart from the significant impact of the project in creating awareness and reducing child labor, the long-term advocacy has resulted in Gujarat Government mandating the formation of child protection committees in all villages that work with local panchayats, to ensure upgrading Aanganwadis and local school infrastructure.

Another unique child education programme that is commendable is their work with construction labourer's children, in their Child Friendly Spaces, a neglected area.

SAATH has done commendable work in the area of children's education and women and health and should be congratulated on several counts.

In the future, they must review their vision, strategy, and mission on the impact they have had, in order to achieve an equal and just society. Continuing a "welfarist approach" may not work as the inequality is increasing and governments are already taking cognizance of the fact that it needs to ensure quality education of marginalized communities and the health and well being of all (Ayushman Bharat 2018).

In such a scenario merely addressing the practical gender needs of providing better access to water, sanitation, supplementary food, pre-school education, livelihoods may not be the most strategic way to proceed. NGOs / CBOs need to start working on the strategic gender interests of women and girls, which will lead to a more equal society.

"Real Empowerment" demands more than livelihoods provision. It includes entitlement education, which questions the social and cultural norms. It requires communities to ensure violence-free safe spaces where women and children feel safe to learn, earn, and live optimal lives.

With three decades of experience, SAATH is uniquely poised to take a quantum leap to make this happen. In addition, there is merit in SAATH questioning and addressing caste / class and religion barriers, the brunt of which is majorly experienced by women and girls as they are expected to be the cultural ambassadors of the families and communities.

Amitabh Behar, CEO, Oxfam India and Former Executive Director of National Foundation for India

I grew up with a belief that 'India lives in its villages'. This was taught to us in schools, colleges and was part of the public narrative. The entire imagination of working in the development or voluntary sector was the romance of going to remote rural areas and work in difficult conditions for poverty alleviation and/or rural development. To challenge this understanding that for meaningful social action, one does not need to go to rural areas; and to acknowledge that powerful social work is possible and needed in urban areas was a radical idea three decades ago. Even today, the developmental civil society working on urban issues and with urban communities are very few. Many voluntary sector organizations and NGOs are now based in urban centers but they continue to do work in rural areas or on larger macro issues through advocacy and research. This ability to challenge the established and settled narrative of development and change, requires three important characteristics. First, is the ability to think out of the box or be irreverent and innovative in analyzing the world around you. Second, ensure actions and understanding which is dynamic and responsive. Third, deep commitment to certain core human values, and resolve to defend them in adverse conditions, even if one needs to fight a lonely battle. **Saath was born with this ability to break new ground, in terms of pioneering and fresh work; along with these three important features; it has consistently displayed these qualities in its three decades of work.**

Saath embodies each of these characteristics abundantly. Infact, the last thirty years' journey of Saath stands testimony to these remarkable and exceptional qualities. The ability to live and practice these qualities is evident in the journey of the organization and its ever evolving agenda in response to the changing socio-economic and political milieu. All these qualities coalesce around the central idea of solidarity – togetherness or more aptly called 'Saath'. I must say that these observations, though expressed in abstract are grounded in a serious reading of the solid process documentation of Saath's work over the years. My reading of the work of Saath makes me feel that the most important and valuable characteristic of the organization is to ensure invisible 'saath'. The operative (and inspiring) word for me in this context would be 'invisible'. The greatest tribute to an organization is its ability to be a supporter, facilitator or a catalyst and not be the face of change. And it's a bigger tribute when the organization has the ability to be invisible and gradually fade out; and through this process the communities get empowered and build agency to lead their own struggle for dignity and justice.

I am sure each piece of Saath's work would stand testimony to the core principles of Saath's ways of working and belief system. **Their work with young people is a fascinating journey of an organization learning to work with communities and of providing leadership in a way that it never crosses the boundary of an enabler, while ensuring that the young people are able to channelize their confusions, anger and passion for change and better life into building pathways for the entire community towards a dignified life.** This is a short note so a detailed analysis of the youth work of Saath is not possible (which might be a book in itself), but let me point out three phases which I see as distinct milestones. The first is starting work with young people, particularly through sports (volleyball) and building trust and rapport with the young people, which became one of the mainstays of Saath over the decades. Second, milestone is the decision of Saath (almost intuitive but extremely courageous) to not shy away and respond to the hostile and difficult environment of communally tense Ahmedabad (post the riots in 2002) again by working with young people for building abiding peace. The third phase is the realization and recognition Saath made towards the end of the first decade of the new millennium that the young people are lost and feeling diffident in a neo-liberal market led world where life opportunities have remained largely outside the reach of basti youth (the investment of Saath in livelihood in a big way). In all these phases, the story of Saath is similar and therefore more amazing. The adherence to a core values and ethics framework guided

Saath in different phases and different context, interestingly to the same result – success in finding solutions to the issues of the community (as co-travelers and not as solution providing experts). At the core of all the work of Saath has been the firm belief that the people, particularly communities and groups who live on the margins, have to be at the center of all socio-economic and political processes for lasting and dignified change. The process of change itself should be empowering and not an outcome delivered by outside interventions.

As we are all aware that we have already moved into the phase of demographic dividends for India, but unfortunately this is just a slogan with serious thinking of how this enormous advantage of a young population can be harnessed for development of the country and its people. Infact, many have already started sharing the concern that this demographic advantage can lead to a demographic catastrophe if we do not engage in channelizing the youth energies and capacities in the right direction. An organization, like Saath, needs to play this historically important role, given its experience and commitment to working with young people. It needs to move very quickly in building a comprehensive response along with fellow organizations, to address this question of how do young people shape the future of India (and in doing so what is the support they would need)? The support and accompaniment needed should be the big agenda for Saath in the coming years.

Prof Ram Kumar, IIMA alumni who teaches marketing at premier management institutes and conducts Corporate Training Programs in service/knowledge-based industries

The Saath organizational ecosystem has clearly identified a new "market" and is addressing its needs with different initiatives

- The phenomenon of
    - Migration to urban areas
    - Urban poor
    - sub-optimal infrastructure for these migrants has been identified not as a problem but something that needs to

be taken in stride and also creates huge opportunites for National Growth
  ◆ Indeed this approach can be translated to countries across the globe for while the magnitudes may be different, the stories are the same.

- Saath and its constituent programs also address the urban poor in a completely different way – not as a problem to be solved – but as an integral component of modern economic growth. As so eloquently stated in the introduction an urban resident needs a plethora of services to run her personal and professional life:

  ◆ gardener
  ◆ Driver
  ◆ Plumber
  ◆ Carpenter
  ◆ Security guard
  ◆ Home assistant
  ◆ Child care
  ◆ Vegetable vendor
  ◆ newspaper/milk etc delivery personal

  It is important to note that these jobs will NEVER be done by the people who populate the 'haves' strata. They will have to be performed by the irabn poor

- To cater to such a 'have' market Saath concentrates a lot in training and skill development: Home Manager, Udaan, Youth force…are all examples.
- Importantly, and very innovatively Saath's programmes repeatedly assert, process and deliver that the urban poor and the superb goods and services they deliver are not "marginal, informal or unorganised". **If they can power Urban India when branded like this imagine how much more they could contribute to the national economy if empowered.**

- Saath's market facing initiatives also take into account that the kinds of services the urban poor deliver are not what they want their children to do. They want their children to study, become graduates and live a better life. Unfortunately, this will only add more numbers to the existing problems the "haves' face: unemployment, underemployment, job-dissatisfaction, lifestyle diseases etc etc. Converting the "marginal, informal or unorganised" sector into a legitimate, productive and totally mainstream set of vocations and avocations will substantially change the urban landscape.

Prof. (Dr.) C N Ray, former professor at the Faculty of Planning and Public Policy, CEPT University, Ahmedabad

Gujarat is susceptible to various types of natural disasters. The January 2001, earthquake devastated several cities and villages in Gujarat. The epicenter was near Bachhau town in Kutchchh. Along with Kutchchh, the impact of the earthquake was seen in several other districts in Saurashtra and central Gujarat. In Ahmedabad city, many high buildings collapsed or developed cracks killing over 700 people and destroying more than 100 buildings. The Kutch district is an arid region, vulnerable not only to earthquakes, but to other natural disasters like cyclones and droughts. The earthquake struck in the wake of two consecutive years of drought in 1999 and 2000. In most of the affected districts, the main sources of employment are agriculture, animal husbandry, salt mining and refining, handicrafts and trade, which were highly affected.

NGOs and civil society groups from across the country played an important role in the relief and rehabilitation work. The quick adoption of different policies paved the way for participation of community groups and representative institutions in the decision-making process. This helped in reflecting people's priorities and aspirations in the programme deliverables. This strategy was intended to help the affected people and stakeholders to make informed choices regarding their habitat by disseminating knowledge on seismic technology, building materials, and construction practices. The guidelines issued by the state agencies encouraged the participation of many private sector companies, NGOs, and expert institutions in the post-earthquake relief and rehabilitation work.

After the earthquake, Saath was involved in relief work in 17 villages of Khadir and Rapar town. Its involvement started with the task of carrying out a rapid assessment of all damage caused in the villages. It remained active in different short- and long-term rehabilitation work, as an active partner of a NGO network that played a very important role to bring normalcy in the affected villages and towns.

The learnings from relief and rehabilitation of the Gujarat Earthquake shows that the active participation of NGOs and civil society groups along with state agencies helped in providing relief to earthquake affected people in Gujarat. However, one can easily see that places with certain communities known for their inward-looking mindset, delayed the process of post disaster rehabilitation work. It appears that existing groups were rigidly divided into two main caste-oriented formations. The division had also manifested itself by these groups' identification with different political parties. The caste and political divisions affected almost all aspects of economic activities in Khadir. It is also realized that very little work with women was possible because of the gender rigidities and lack of exposure to developmental activities, compared to similar rural areas elsewhere in Gujarat. In contrast, working in some districts were less stressful due to more interaction between differing castes and communities, which helped in undertaking post-disaster relief and rehabilitation work.

Saath now has enough experience of effective participation in post disaster relief and rehabilitation work. It can easily go beyond the state boundaries. The method it has followed is quite interesting based on the principle of local participation, and mobilization of affected communities. It is a very useful learning.

It is important to note the process of how access and delivery of design, materials, skills, technology, and finance takes place. Livelihood, as well as housing, is a dynamic process when linked with other requirements and considerations.

The post-earthquake redevelopment process shows very clearly that empowerment is the core of participation and it can be achieved through conscientisation. Participating agencies should work and ensure that participatory processes get institutionalised at the local level for future development.

Agencies like Saath can play a facilitating role in terms of providing advice on livelihood, restoration of affected buildings and infrastructure facilities. The disaster affected families can be mobilized to make a choice based on individual needs and requirements. Creating such an environment will encourage local level participation and empower the community to undertake activities on their own. Saath should also ensure that it could work within a large public agency framework without compromising its own philosophy and methods of working at the local level. Staff and volunteers with a proven commitment to work along with community groups will certainly help to ensure that it remains a development institution, rooted in participatory principles of local level planning and development.

As the country remains highly vulnerable to natural and man-made disasters, there is need to develop the capacity of community groups, agencies and institutions at various levels. NGOs like Saath can undertake the task of creating awareness among the communities and the administration. There is a need for a continuous initiative for community-based training, establishing institutions, creating and strengthening of disaster management organizations, for capacity building, to cope with such disasters. It will also reduce heavy dependence on international organisations and associated experts in developing countries.

In 2015, the Sendai Framework for Disaster Risk Reduction emphasized building the Resilience of nations and communities to disasters by using the experience gained through the regional and national strategies or, institutions and plans for disaster risk reduction. Working in post-earthquake relief and rehabilitation has made Saath adopt a concise, focused, forward-looking, and action-oriented framework.

Now it is recognized that capacity building in disaster management is very important for prevention and reduction of loss of life and property in all kinds of disasters. The capacity building can act as a method that will enhance competency of community, public institutions, and all other stakeholders. **Saath can help in developing awareness, conducting training and play the role of a proactive disaster management institution, not only for Gujarat, but for the entire country.** It will benefit future policy formulation in disaster management that will minimize loss of life and property.

Prof Amita Bhide, Dean of School of Habitat Studies, Tata Institute of Social Sciences

Revitalising Community Development

Community Development: Relevant anymore?

Community development is an idea that developed in the 1940s through several experiments in various parts of the world. This included work in developed countries but also includes experiments such as Sri Niketan led by Rabindranath Tagore in rural areas of Bengal, trying to combine social transformations with efforts to marry tradition and new science for scripting a vision for modern India. When the country became independent, community development was the strategy of choice for change in villages ushered by gram sevaks who would bring knowledge produced in universities to the villagers and supported by extension departments.

In the 1960s, there were efforts to professionalise community development as a method of social work. Prime among these is the definition of Murry Ross who famously described community development as the process by which 'a community identifies its needs/problems, prioritises the needs, identifies resources to resolve the problems, undertakes action vis-vis these needs and in so doing develops the collaborative and other capacities (Ross, 1967). The Gulbenkian Foundation played a critical role in formalising and privileging this approach by preparing a report that brought together prevalent knowledge about the practice of community development (Gulbenkian

Foundation, 1968). Community development was thus a part of the promise of the new world post the second world war where it was seen as a bridge between the state and people; between scientific, modern knowledge and technology and people embedded in tradition; and between various groups of people divided by caste, class and privilege. The bridge was to be constructed by the efforts of professional social workers.

Yet by the 1970s, community development was given up as an approach. The shift away from community development was not just seen in the shift of development policies but also in the professional social work realm where the consensual approach of community development gradually gave way to the more political and activism oriented community organization. Community development was critiqued for assuming that structural conflicts and power relations in village and other communities could be transformed into consensus and that benefits of development could be shared equitably by all members of the community. In most contemporary literature, community development is regarded as a thing of the past.

Saath and Community Development

Saath in Ahmedabad is working with one of the most contemporary challenges faced in India – the issue of informal settlements or slums. Slums embody the inequities in urban societies; but they are not homogenous. The contexts of emergence of these settlements, their ability to consolidate, access basic services and stake their rights to city are mediated by regions of origins, caste, religion; ownership and legal status of land which they occupy and the consequent vulnerability to eviction and the overall approach of the state and land markets to these settlements.

The challenge that Saath set for itself in creating a secure, liveable space for these settlements in an expanding and gentrifying city is thus an uphill battle. The challenge is made even more difficult due to the nature of politics generated around slums. The experience of electoral and other politics creates a trust deficit among the inhabitants. They are wary of outsiders but also of the possibilities of change. Settlements like Juhapura, which have faced traumatic experiences such as riots and exclusion, embody this mind set at an extreme level.

It is in this context that Saath decided to adopt a community and integrated slum development approach. What did this mean? It implied a consistent belief in the capacities of youth, women and leaders from the community; an investment in a process of bringing these sections together and giving them an opportunity to know each other as fellow travellers, beginning to aspire and becoming partners in an uncertain though steadfast path to changing their circumstances. Change in slums requires the support of the powerful stakeholders in the city such as the municipal corporation and the corporates, which was mobilized by Saath. However, the organization also ensured that the youth, women were the initiators, implementers and partners of this change. This ensured the sustainability of the change and reiterated the fact that ultimately all stories of collective change are also stories of individual triumph against adversity. Facilitating such change among disenfranchised people who dare not become a community is not easy, the chapter on community development thus hides more than it reveals. Yet it is indeed inspiring to see that people from Gupta Nagar, Juhapura and other slum settlements are not just able to undertake material and infrastructural change in their living conditions, but also able to build bonds within and across communities.

What lies ahead

Does the experience of Saath mean that one can revisit the community development approach? One key difference in earlier community development approaches and the one adopted by Saath is the proactive acknowledgement of conflicts and power relations in the settlements. Thus, community is not taken for granted or assumed to be; it becomes a project to be worked at. In fact, the experience of women in Gupta Nagar illustrates that the conflict between women and men became an opportunity to assert the collective capacities of women to renegotiate community power relations.

Another feature is the sequencing of change projects that move from easier ones to those that are more complex and strategic i.e. those that involve embedded patterns of exclusion. Infrastructural projects often create opportunities of individual incentives, which can undermine the collective process. However, the strong base of collective forums can help to counter and balance the individual incentives.

Another critical facet of community development approach of Saath is the time and effort invested in initial trust building with the settlements and the consciousness over its own 'outsider' status. Several organizations working in informal settlements often cross this thin line and consider themselves representatives of the community, perpetuating vested interests at another scale. Saath however has used this 'outsider' status as a strategic choice to enable capacity development of the local groups.

These features represent a reinterpretation and hence revitalization of the community development approach with a great potential to solve some of the most pressing contemporary challenges.

Prof Howard Spodek – Eminent Historian and author of "Ahmedabad: Shock City of Twentieth-Century India"

Saath and the World of Indian NGOs

India is one of the world's most lively homes for NGOs with over two million of them, an estimated one for every 600 citizens. They come in all sizes, from a few hundred members to more than a million; all forms, from local institutions to those with national and even international scope; and all missions, from assistance to the poor, to medical assistance, to environmental issues, to low cost housing, to citizens' rights.[16]

---

[16] Times of India February 23, 2014. https://timesofindia.indiatimes.com/india/India-witnessing-NGO-boom-there-is-1-for-every-600-people/articleshow/30871406.cms

Recent Problems in Government – NGO Relations

Many NGOs in India engage in social services that in other parts of the world are performed by the government itself. To some degree, governments – central, state, and local – engaged the services of NGOs in social service delivery because the government's own financial resources and administrative capacities were limited.

As governments' resources have increased, however, they have increasingly provided these services on their own, thus boosting their prestige among the population and building their patronage networks among their expanding pool of employees. Conversely, NGO visibility, capacity, and employment have been diminished.

Perhaps even more challenging to NGOs in recent years, many politicians see them as threats. They see at least some NGOs mobilizing the citizenry to demand more civil rights, or more services, or more attention to their civic concerns. Some of these goals may challenge government's own wishes or powers.

To the extent that NGOs receive foreign funding, they are also seen as serving the interests of their funders rather than the interests of India, and, often more to the point, as supporting causes challenging or opposed to those of the government, or at least of the ruling party, or of the dominant religion.

Many governments, at all levels, therefore seek to limit the independence of NGOs. For example, in April 2015, the Government of India shared a list of over 42,000 NGOs with Financial Intelligence Unit (FIU) to check suspicious foreign funding amid the crackdown on some top international donors for flouting the Foreign Contribution Regulation Act (FCRA), 2011. Intelligence reports claimed that several of these organisations were diverting funds for purposes other than the permitted use of foreign contribution.

To some degree the criticism of foreign funding appears valid, at least to some extent. There is an accountability deficit among the NGOs. Many don't submit details of receipt of grant and spending to income tax authorities. The funds are substantial. For the financial year 2010–11, available data show that about 22,000 NGOs received a total of more than $2 billion from abroad, of which $650 million came from the US.

SAATH, however, is an outstanding example of an NGO that benefits the society without challenging the government, and often by working side-by-side with it. SAATH is not confrontational. Many of the services it delivers include participation of the private sector in a relationship that is market based.

SAATH's breakthrough initiative began with the slum networking program, about 1995. The usual solution for slum problems called for evicting the slum dwellers and razing their homes. SAATH, however, had learned from the social entrepreneur Himanshu Parikh of his successful "slum networking program" in Indore. SAATH put forward a similar proposal to the Ahmedabad Municipal Corporation (AMC): upgrade the slums; improve them without moving them or destroying them. Make them clean and livable. Keep the slum community in place, but improve its physical conditions. Slum residents would pay part of the costs for upgrades and regularization of services—water, roads, streetlights, and proper toilets. The AMC accepted the proposal.

The program was a risk for the young NGO; for the city government; for the industrialist, who helped to sponsor it; and for the people of the slum in which it was introduced. But it worked and became the model for the "Slum networking program" (SNP) in Ahmedabad that ultimately touched 9,348 families.

In collaboration with USAID, SAATH extended this principle of beneficiaries paying a substantial proportion of the costs of the services they were gaining with a program for bringing electricity to the slums. Building on this model, the program was taken forward by Torrent Electricity Company itself. Today, almost all slum households in Ahmedabad are electrified and connection charges have been slashed by half. Company revenues have shot up by 30 to 40 percent with 200,000 households paying for the services. Here again SAATH established a collaborative pattern of work that included the NGO, private industry, and the beneficiaries.

Underlying the problem of slums is the problem of poverty and unemployment. Part of this problem might be solved if job training might be more closely linked to industry needs. SAATH consulted with industry experts concerning their needs and then began training programs to increase the job skills of semi-skilled workers in the informal sector and to train youth from underprivileged families for employment in the formal sector. It provided micro-entrepreneurship and vocational skilling of women in non-traditional trades. It stirred interest in education among adolescent girls by imparting activity – based learning in science, technology, engineering, and math, subjects often neglected in the education of girls.

As India was becoming both more urban and less poor, SAATH saw new needs and new opportunities. Two of its ventures were particularly innovative. SAATH realized that some construction companies were assessing the benefits and risks of building for poorer citizens – not the poorest, but those who could for the first time consider buying their own apartments, those who were entering the lower middle class. Meanwhile the potential buyers also had to assess their benefits and risks in entering the housing market as consumers. SAATH joined with a construction company DBS Communities and established strategies for enabling potential buyers and sellers to strategize appropriately. SAATH was successful in this because it could earn the trust of both builders and buyers

Construction was becoming India's third largest economic sector behind farming and industry. Construction workers on large projects are migratory; they follow the geographical needs of the builders, moving periodically from site to site. As a result, their children often do not receive formal education, or even adequate housing, as their parents are on the move. In another new initiative, SAATH undertook to provide basic education and also access to water and sanitation for the children of these migratory workers in and around Ahmedabad. The funding often came from the CSR (corporate social responsibility) funds of major, highly-regarded construction companies, like Godrej and Bakeri.

Metro and Urban Resource Centers

Expanding cities also build increasingly large and complex transportation systems. In the process, they displace large numbers of residents whose homes stand in the way. The Ahmedabad Metro-rail system – MEGA – needed a third party to clear the way for its construction by relocating citizens. MEGA called for bids – and, in recognition of its track record, chose SAATH.

SAATH's latest innovation is the Urban Resource Center (URC) which serves as a critical linkage point to connect slum residents with information, knowledge, and services from governments, NGOs, and private companies. To date, four URCs serve 13,000 households across Ahmedabad for an annual subscription fee. SAATH is also actively working with academic institutions, government, and other NGOs to develop policies for coping with critical problems of urban slums.

Institutional Style

As its contributions increase, SAATH's style of work demonstrates the value of this exemplary NGO:

- It has partnered with others – other NGOs, governments, academic researchers, private businesses – in order to achieve mutual goals.
- It has been nimble in seeing what works and what does not, which projects are rising in importance and which are no longer viable and it sets its priorities accordingly.
- It has embraced the "vision of an equitable city." Its focus has been almost completely urban. As many other NGOs that were focused on rural areas and issues begin to shift toward cities, SAATH serves as a model.
- It has seen the importance of including market realities in its undertakings.

SAATH has developed international liaisons, and not only with NGOs or funding agencies overseas, but also with students. Each year, SAATH hosts groups of university students, usually with their professors. The benefit is mutual: the students and professors learn a great deal about ground realities and problems in India; the hosts get help with their ongoing fieldwork; and sometimes useful suggestions for improvements also emerge.[17]

SAATH has empowered its own staff and second line of command as well as the beneficiaries of its programs. As founder-director Rajendra Joshi remarks, "It is a fun place to work, supportive and empowering for its staff, committed to capacity building."

In its practical accomplishments and in its exemplary style of work SAATH has demonstrated that, despite the apprehensions, and occasional hostility, of various levels of government in India, NGOs have a distinctive and productive role to play in the development of the country and its people.

---

[17] I have witnessed this kind of creative interaction between SAATH and a group of students from Duke University, USA, in the summer of 2017.

ANNEXURES

Annexure 1

Social Return on Investment Study on Saath conducted by 4th Wheel

Introduction

Social Return on Investment (SROI) is a framework for measuring and accounting for the value created or destroyed by our activities – where the concept of value is much broader than that which can be captured by market prices. SROI seeks to reduce inequality and environmental degradation and improve wellbeing by taking account of this broader value (The Seven Principles of SROI, The SROI Network).

Originating from the financial ratio of return on investment, the concept of social return on investment (SROI) is a method for understanding, measuring and reporting on the holistic value that is created. The process of an SROI analysis leads to what is called the SROI ratio. This is the ratio between the value of the benefits and the value of the investment. For example, a ratio of 1:1.5 indicates that for every Rs. 1 invested in an activity, project or program, Rs. 1.5 of value (economic, social and/or environmental) is generated for society, generating an added value of 50 percent.

SROI bases the assessment of value on financial returns as appreciated by the perception and experience of the stakeholders themselves. It looks for key indicators of what has changed, tells the story of the change and, uses monetary values for these indicators. By monetizing these indicators, financial equivalents to social and environmental returns are created. This allows combining the created outcomes and expressing them in one common value that can be understood by actors outside the sector as well.

To evaluate the impact of Saath's work in the last thirty years, the SROI methodology was adopted, to measures change in ways that are relevant to the people or organizations that experienced or contributed to it. The analysis helped stakeholders to value, reflect on and measure the results of activities in a participatory way, thus enriching the (development) objectives of the organization and providing learning opportunities.

Scope of the Evaluation

The study undertook a Retrospective SROI Evaluation, based on actual outcomes that had already taken place, owing to SAATH's interventions. Although SAATH has impacted over a million lives, the evaluation assessed the value created in three main slums of Ahmedabad that SAATH has worked in, over the last thirty years. These include slums of Behrampura, Vasna and Juhapura which impacted 21,653 persons.

The crux of SAATH's work lies in the Integrated Slum Development (ISD) Approach, which strengthens the urban planning process by integrating the urban poor in the city planning and development process in a participatory manner. The aim is to create a healthy and enabling urban environment by providing adequate shelter and basic infrastructure facilities to slum dwellers. Activities and programs of the ISD framework are designed to enable slum residents to realize their potential by increasing their management and technical expertise, nurturing leadership, and enhancing their self-esteem and self-confidence. These abilities start a virtuous dynamic cycle of actions for development, that would lead to improvement in quality of life, that would enhance their self-worth, which would lead to critical reflection and finally to more actions for development.

Interventions under Study

1. Education for Pre-school Children

The intervention focused on building preschools for urban poor that address the basic growth and development needs of children like: basic education, health, nutritional and recreation. Children between the age group of 3 to 5 years were the main beneficiaries of the program.

2. Mother and Child Care

Activities under this domain focused on both, health and education outcomes. Mothers were educated and supported to improve their child's education status, nutrition and care. Mothers living in slums were direct beneficiaries, while children also experienced value from the program.

3. Tuberculosis Prevention

Improving access to healthcare facilities and curtailing the spread of the disease were the main ambitions of the program. The program broadly comprised of awareness sessions and distribution of nutritious food supplies. Existing and potential patients were also mobilized for diagnostic tests. Patients were supported and motivated to seek and/or complete treatment.

4. Youth Force

The program formed youth groups where youth collaborated to discuss issues of their neighbourhood, plan and execute solutions and activities, and organize community level activities. Themes mainly revolved around environment, self-identity, sports, financial inclusion, health, and skill development and government schemes.

5. Youth Employment

Saath provided vocational training to youth between 18 – 35 years, in the areas of computer education, personal care and beauty parlour services, retail management etc. and placed them in jobs. Youth were trained in functional job requirements, based on market needs. Saath also enrolled and trained workers in the informal sector by providing them vocational training in the areas of electrical work, carpentry, masonry and plumbing so that they can increase their earning potential.

6. Women Livelihoods (Home Managers)

This program was developed with an objective to build professional identities of marginalized women working as domestic help in Ahmedabad. Saath provided a structured training pertaining to home management in various domains like cooking, cleaning, baby-sitting and geriatric care.

7. Urban Livelihoods and Entrepreneurship

The program aimed to encourage setting up new nano-enterprises, and provide support to existing entrepreneurs. It offered education on theoretical concepts and practical learning to new and existing entrepreneurs. Mentors help enrolled participants to understand various facets of entrepreneurship, apply the learning and constantly assess the impact the program has had on them.

8. Incremental Housing/Basic Services

The major components involved were separate household connectionss for water supply, household toilets, drainage system, pavement for internal roads, streetlight, landscaping, and facilities for solid waste management. Saath helped in facilitating payment and credit for the residents' contribution towards the project.

9. New Affordable Housing

Saath promoted a holistic housing rights approach, which enabled socioeconomic equity and development. This programme created awareness amongst the slum communities about their housing rights and organized them again any illegal evictions which were carried out without prior notices and undesirable resettlements plans.

10. Electricity Supply

The pilot project provided 1,000 household level electricity connections at Rs. 2,000 per connection. The project had two objectives, first to check if residents would pay Rs. 2,000 for a connection and, if the monthly bill would be affordable. Moreover, an electricity bill was a very important document as proof of ownership; which was made available to them.

11. Financial Inclusion

Savings groups were formed for these communities which were then registered as savings and credit cooperative societies. The savings and credit programme then evolved into a professional micro finance basis, with more than 3,500 members. The Joint Liability Group (JLG) practice was then adopted.

12. Urban Resource Centre/Citizenship

The URC had volunteers that counselled people regarding their needs and even accompanied them to the respective civic bodies for enrolment/application for required government identities or public benefit schemes. This equipped marginalized communities to avail of government documents and schemes.

Stakeholder Analysis

Based on identified interventions, an in-depth stakeholder mapping was undertaken with Saath's team and all stakeholder groups, who experienced change as a result of activities of the project and have invested in the projects. (The list of identified stakeholders is attached as Annexe 1). Key stakeholders were women, youth and community members (mostly adult male) for inclusion in the study, based on them being the main beneficiaries and target participants of the programs which were implemented.

In addition, all Saath's implementation teams and experts working on slum and urban development were included in the study. Semi-structured interviews were conducted with them to gain deeper insights into the cultural nuances of and value created in communities.

Sampling

Calculated for a sample universe of 21,653 in three slums, assuming 95 percent confidence level and 5 percent margin of error, 377 people were aimed at, to be included in the study.

Area	Respondent Size
Behram pura	79
Vasna	183
Juhapura	155
Total	**377**

Each slum was looked at independently to ascertain the sampling frame. Since programs spanned thirty years, beneficiary reach was broken down to six distinct phases. Proportion to percentage sampling was then applied where beneficiaries in each program across the six phases were looked at and the final sampling was determined. The final number of respondents (women, youth, community members) were then identified using stratified sampling, according to age, gender, involvement in and type of programs accessed.

Stakeholder Group	Total group size	Number engaged	Intended/Unintended Changes
Women	10,612	185	- Financial Inclusion/Access to Credit - Employability - Health, Safety and Wellness - Children's education
Youth	720	11	- Employability - Leadership, Confidence
Community	9283	162	- Access to basic services. - Access to information - Affordable Housing Facilities - Improvement in Health, Safety and Wellness

Research Tools

Focus Group Discussions

Several FGDs were undertaken with women, youth and community members to ascertain outcomes of the program as experienced by them. Discussions assessed their role and involvement in the program, perceptions of impact, and contextual realities related to integrated slum development.

Contingency Valuation

This method helped to estimate the value that respondents placed on services offered by Saath. The approach asked people to directly report their willingness to pay (WTP) to obtain that service, or willingness to accept (WTA) to give up that service.

Value game

Value Game uses cards with images and short descriptions of alternative goods, experiences and the like, that are tailored to the users of the game. They are then asked to rank and rate these cards so that comparative values of outcomes can be calculated.

Hypothetical budgets

The Group was given a mock budget of Rs. 10,000 to divide amongst all the interventions undertaken by Saath. This helped to understand their view points in terms of % allocation of budgets to various activities and the value they place on these.

Semi-structured interviews

Interviews assessed respondent engagement with the organization and program awareness on the program and perceptions on relevance, effectiveness, impact and sustainability of interventions. They helped to determine important aspects of SROI calculations like attribution, drop-off and deadweight.

Limitations of the Study

Within the SROI framework, abstract, narrative or non-quantitative indicators such as confidence, joyfulness and family relationships have to be priced and adjusted according to stakeholder's perspectives and adjusting factors such as deadweight and drop-off. We have to emphasize that the values are derived from the perceptions of the stakeholders and not from traditional models of predictive financial analysis. The social return on investment figures of this project should not be compared with that of different project because the perceptions and resulting benefits to the beneficiaries of different projects will vary. To make the analysis transparent, we prepare the SROI report in a comprehensive manner and openly explain the outcomes as well as the processes of calculation and derivation. We also list various assumptions and sensitivity analysis used in the process. The evaluation scope spanned a very long timeline, where there is a possibility of value significantly decreasing or increasing and is difficult to estimate. Recall bias was a prominent limitation since some interventions took place several years ago. Among the communities under study, monetizing impact of interventions was challenging. While they were able to articulate both and tangible and intangible outcomes for different activities, giving a monetary value to outcomes was many times random estimates or just 'guess work'. The interconnectedness of interventions made it difficult to isolate monetary values for specific interventions and related outcomes. Spread out geographies and migration of people in and out of the areas, owing to upward social mobility made respondent/sample inclusion difficult.

Mapping inputs, output and outcomes

SROI is an outcomes-based measurement tool. It is important to map outcomes to develop an impact map, also called a theory of change or logic model, which shows the relationship between inputs, outputs and outcomes.

Mapping Inputs

Inputs are the financial value of the investment. Inputs are what stakeholders are contributing in order to make the activity possible and are used up in the course of the activity – money or time, for example. For most stakeholders, the inputs were Effort, Skills, Resources, Monetary contributions and Time.

Mapping outputs and indicators

Through the different research tools stakeholders identified a wide range of outputs. The outputs further were mapped to different indicators. Different data collection methods were used with the stakeholders to also triangulate the data received.

Stakeholder	Output	Indicator	Data Source
Women	Women participating in work, Improved reading habits, Monetary security, ability to take loans, Increase in savings	Number of first time women employees/ entrepreneurs, Improvement in household income with higher earning women households, Higher savings	FGDs, Value Game, Experts
	Improvement in knowledge about health issues	Reduction in number of visits to the hospital, Reduction in TB Cases	FGDs, Value Game, Experts
	Women participating in work, monetary contribution	Number of first-time women employees/ entrepreneurs, Improvement in household income with higher earning women households, Higher savings	FGDs, Value Game, Experts
Youth	High skills among youth, Less dropouts, More education	Increased opportunities in higher skilled jobs	FGDs, Value Game
	Youth Force initiation, Community peace and unity	Increased numbers of youth attend schooling decreased drop out, increased enrolments in higher studies.	FGDs, Value Game
Children	Enrolment of children in preschool, Enrolment through RTE	Increased number of children in Balwadi's, Better nutrition	FGDs, Value Game (Of mothers)
Community	Access to basic facilities (Electricity, Water, Sanitation, Roads), Better living conditions, Better housing facilities	Increased Electricity connections and upgradation of housing	FGDs, Value Game
	Ease to get information and required documentation processes, More information about rights	Increase in documents	FGDs, Value Game
	Housing facilities, Lower interest rates, Increase in quality of life	Number of families who bought affordable houses	FGDs, Value Game
	Improvement in knowledge about health issues	Reduction in number of visits to the hospital, Reduction in TB Cases	FGDs, Value Game, Experts

Mapping Outcomes

The stakeholders had given multiple outcomes of the intervention. Outcomes as attached to each stakeholder are qualitatively outlined below:

Children

Children who have participated in the programs have improved health due to the nutrition support provided. They have academically coped better in schools because of a strong cognitive foundation. Immunization has prevented various illness as they grew.

Women

Women have been empowered in many ways. Timely immunization, hospital deliveries and better nutrition has improved their health. They have improved livelihoods and more income. Their world view has changed. They can take better care of their families. Many of them have become leaders in their communities.

Youth

The youth Saath has worked with have become motivated and self-assured. They have realized that they can bring about change in their families and communities. They can deal with other stake-holders in the city more confidently. They have become role-models and leaders.

Communities

Communities on the whole, have been empowered. They can organize themselves for getting their rights. They do not see themselves as disadvantaged and dependent on welfare support. Their dependency on politicians who considered them as vote banks has decreased. They are more secure about not getting evicted.

Electricity, water, sanitation and housing

Getting access to these basic services has enhanced the physical quality of life and have a multiplier effect. The impact has been reduced sickness and expenditure on health and increased expenditure on education of children. Women do not have to stand for hours collecting water and can use this time productively. The self-esteem of individuals and families has increased because they no longer see themselves as people living in slums. The social status of households has gone up because they live in decent housing.

Livelihoods

Apart from the quantitative increase in incomes, the ability to earn for oneself and contribute towards the well-being of the family has a huge psychological impact. The self-image of being a productive person and the respect that it evinces amongst families, peers and society is considerable. They become contributors to the national economy.

Financial inclusion

Financial inclusion is the backbone of economic empowerment. We closely observed how the ability to bank, save, get credit at reasonable rates, get insurance has contributed to the social and economic stability of individuals and families.

Quality of life

People and communities have seen a considerable increase in the quality of their lives. They live in better and safer physical surroundings. Their incomes have increased, and they are able to generate surpluses. They are healthier and better educated. They use their surplus income to avail basic necessities, purchase vehicles and household appliances. They can now get housing loans.

Citizenship

People feel that they are equal citizens. They now have documents through which they can access their rights and benefits as citizens. They can engage with the government, middle class, private sector and civil society institutions confidently.

Policy advocacy and replication

A number of innovations, programmes and projects developed by Saath have been replicated. The Slum Networking Project was taken up by the Ahmedabad Municipal Corporation at the city level. The Slum Electrification Project was scaled up by Torrent Power in Ahmedabad and Surat cities. Youth employability training was scaled up by the Gujarat government across the state. Other foundations and NGOs have replicated this model. Urban Resource Centres became part of the National Urban Renewal Mission. Social enterprises such as Urmila and Griha Pravesh have been replicated by other organizations. These indicates a much wider social impact of Saath's work.

Financial Proxies

With the help of the value game and hypothetical budgets we were able to infer the common outcomes that were experienced by maximum stakeholders in all the three areas. These outcomes were then further given financial proxies. All values in SROI are subjective and are based on the perception of the stakeholders.

Values are a way of presenting the change that the stakeholders experienced and the financial value that they attribute to it. For outcomes for interventions like providing basic services, micro finance, health interventions identifying the value is relatively easy. But SROI also gives values to actives like reduction in student drop outs which are difficult to put a value to otherwise.

Stakeholder	Indicator	Financial Proxy
Women	Number of first-time women employees/entrepreneurs, Improvement in household income with higher earning women households, Higher savings	Average Savings
	Number of first-time women employees/entrepreneurs, Improvement in household income with higher earning women households, Higher savings	Average income of new enterprises
	Reduction in number of visits to the hospital, Reduction in TB Cases	Average cost of healthcare among families and willingness to pay towards healthcare reported by households

Stakeholder	Indicator	Financial Proxy
Youth	Increased opportunities in higher skilled jobs	Average income of new jobs
	Increased numbers of youth attend schooling decreased drop out, increased enrolments in higher studies.	Reduction in number of drop outs, Women perception score
Children	Increased number of children in Balwadi's, Better nutrition	Higher nutrition of children, Women Perception number
Community	Increased Electricity connections and upgradation of housing	Average cost of hiring external help plus set up cost for all basic services
	Increase in documents	Average savings from using government subsidies
	Average savings from using government subsidies	Average cost of housing facilities
	Reduction in number of visits to the hospital, Reduction in TB Cases	Average cost of healthcare among families and willingness to pay towards healthcare reported by households

Establishing Impact

It is assumed that with or without Saath there would have been some aspects of change that would have happened anyway in these areas either through other organizations, government interventions, the community itself.

It is important to calculate aspects which will keep in check that our numbers are not over claiming and will lead to credible results. Aspects such as deadweight, displacement, attribution and drop-off were measured for all interventions. This was again done through a participatory process with the stakeholders. Also as some of the program timeline has been 30 years, appropriate drop offs have been considered.

At the end of this process the impact is calculate with the financial frequencies that we have received and by discounting all 4 aspects.

Calculating the SROI

The social return is expressed as a ratio of present value divided by value of inputs.

SROI ratio = Present Value/Value of inputs

The net social return divides the net present value by the value of the investment.

Net SROI ratio = Net Present Value/Value of inputs

Net Present value is the value in the present of a sum of money, in contrast to some future value it will have when it has been invested at compound interest. Since the programs started as early as 1989 to current 2018 a discount rate of 8% was assumed as an average rate for these 30 years. A discount rate is the minimum interest rate set by a countries federal reserve for lending out to other banks.

SROI was calculated for each of the three slum areas as well as an overall SROI value was calculated. The total financial value of the inputs for the three areas was Rs. 12,87,70,000 and the total present value was Rs. 2,54,92,13,918. This provided a SROI ratio of Rs. 20.80 of social value created for every Rs. 1 of investment. While Behrampura had a ratio of 17.10, Vasna 33.39 and Juhapura 4.24.

Area	Behrampura	Vasna	Juhapura	Overall
Input	3,62,70,000	5,73,50,000	3,51,50,000	12,87,70,000
Total PV	62,01,20,652	1,90,89,79,353	14,88,83,913	2,67,79,83,918
NPV	58,38,50,652	1,85,16,29,353	11,37,33,913	2,54,92,13,918
SROI Ratio	17.10	33.29	4.24	20.80
Net SROI Ratio	16.10	32.29	3.24	19.80

Overall there were similar projects implemented in all areas, the input levels were different, and the stakeholder perception differed from area to area. Since Vasna alone had an intervention on affordable housing, this led to a higher value given by the stakeholders as the price of the houses have also increased over a period of time. Similarly, Saath started working in Juhapura in 2003 where the intervention was limited to pre-school education, financial inclusion and livelihoods

This evaluation applied the Social Return on Investment (SROI) approach, which is commonly used for measuring social impact. 4th Wheel acted as an impartial third party and employed a rigorous and meticulous evaluation process in the implementation of the project, including communicating with stakeholders involved in the project as well as adopting a professional attitude of integrity and accountability, so that the impact of the Saath interventions can be identified in a more objective and comprehensive manner and the benefits of the project maximized.

The SROI analysis demonstrated numerous positive outcomes from Saath's activities. Overall, the various projects had an INR 20.80 social return on investment for every rupee invested. This indicates a 2,080% of value generated for each rupee invested in Saath.

The parameters for measurement have been the number of people who have benefitted, awareness and attitudinal change and, social and economic progress. The effects are both, short and long term. Saath reached 21,653 persons in the three slums to enhance the quality of life of vulnerable people living in low-income neighbourhoods through an empathetic, integrated and collaborative approach. Overall Saath has impacted 15,52,670 persons during the last 30 years.

Communities received significant value from the project, with outcomes related to empowerment, self-esteem and skills contributing highly to the final result, and participants generally thought that these outcomes were life-changing and would stay with them their whole lives. Many factors were seen to contribute to outcomes that stakeholders experienced, by far the most important being their own commitment to taking the program and its benefits forward and applying them through building their lives.

It is important to recognize these extraordinary achievements over the last 30 years, and support Saath to expand its sphere of influence as well as help to raise its social value and attached community impacts.

About 4th Wheel

The 4th Wheel was founded in 2010 with a vision to improve and strengthen social development programs. Over the years they have undertaken various impact evaluation studies with Governments, NGOs, corporates and social enterprises. They focus on periodic, objective assessments of a planned, ongoing, or completed project, program, or policy, to answer specific questions related to design, implementation, and results.

Annexure 2

Reconciliation Study after Gujarat Riots (2004)

Context

Post communal violence of February-March, 2002 in Ahmedabad, Saath has been working towards conflict resolution and peaceful co-existence between communities. Success in development is best measured by peaceful, healthy co-existence, where people, irrespective of caste or community, make space for each other.

Saath started working with the affected slum communities in 2002. Studying and understanding the salient workings of different communities living in close proximity, initiated a process that led to the comprehensive Integrated Settlements Development Programme in these communities. This programme addressed health, education, livelihood, financial inclusion and physical infrastructural needs of the poor and vulnerable.

The process is slow and long, but the right one. Two years after the communal violence, Saath is happy with the ground result.

Visitors (from the field of development) to our work areas have shown a deep interest in the way Saath and the CBOs have been able to successfully translate the vision into reality.

Objectives of the study

- To create a comprehensive database which would also act as a guide for peace building through development initiatives.
- To understand the processes instrumental/detrimental in meeting the needs of development, as well as peace building.

Methodology

This report is based on data collected through qualitative techniques of research, like Focus Group Discussions and Personal Interviews, as well as from Saath's extensive experience.

Two teams, four members each, with representation from the Community Based Organizations (CBOs) and Saath workers from the riot affected areas, were formed.

Focus group discussions would be held with four groups in each riot affected area that Saath was working. These were a Women's group, a Youth group, a Mixed group of males and females and

Committee members of CBOs in the riot affected areas. 25–30 members formed a focus group for relevant discussions. Through discussions, issues to be addressed at group discussions were finalised. Questions specific to the composition of each group, was also finalised.

One member from each of the CBOs that had worked with the riot victims since their days in relief camps – Sakhi Mahila Mandal, Ekta Yuvak Mandal and Sankalp Mitra Mandal, were interviewed. A local resident who had been adversely affected by the riots and lost everything, was also interviewed from each area. These people were from among those who had taken livelihood loans from Saath and were doing well in their businesses, as well as those who were inclined towards working for development in their area. Interviews were also conducted with local workers from each of the riot-affected areas.

The focus group discussions aimed at eliciting opinions on issues like:

- The changing inter personal relationships between the two communities after the riots
- Effect of the riots on the mental and physical health of the people
- Effect of the riots on the economy and livelihood
- Reaction of the youth, women and men and their responses (both negative and positive) to Saath's interventions
- How did the local residents visualise their future and the future of their area with regard to development and peace.

The focus group discussions and interviews with different groups across communities ensured that accurate and unbiased data was elicited for the documentation.

Each team was given two areas to conduct group discussions and personal interviews. The discussions and interviews were recorded and documented. A collative report was prepared from each area, and analyzed.

Background of the Study

Disasters bring devastation and misery, more so, for socially engineered disasters. In this case, it was one human being consciously harming another. Communal riots, one of the most ghastly socially engineered disasters, where, in the name of religion and sect, one human being becomes the enemy of another.

The communal frenzy that completely shook Gujarat and, especially Ahmedabad, in February-March 2002, needs no introduction. Men, women, children lost their lives in large numbers. There was large loss of property too – people lost all that they depended on for their survival.

A large number of people took refuge in the camps that were set up in different parts of the city. When curfew had been imposed in Ahmedabad and the situation was extremely tense and volatile, members of Saath felt that they could not just sit back and observe the situation; they had to reach out to the hapless victims.

Saath had immediately responded to the situation and joined the Citizens' Initiative, a consortium of NGOs, committed to the cause of emergency relief and rehabilitation.

The Saath team met the CBO, Sakhi Mahila Mandal (Sakhi) members. The Sakhi team initially was very apprehensive about helping the riot victims in an environment where fear had gripped every heart in Ahmedabad.

After a lot of deliberations, one of the women Sakhi members came forward saying that she felt it was her moral responsibility to help the riot victims, especially women and children, who had faced such wreckage and sorrow. She convinced a few other members. Subsequently, a team of three decided to visit the camps. Nobody had any idea as to what they were going to do or, how they were going to approach the camp inmates, but each one believed that the need of the hour was to reach out in whatever way they could.

When the Sakhi members visited the camps on the first day, the Muslim community refused to have anything to do with Hindus, even talking to them. They were adamant that they did not want to see any Hindus in the camps. One of them said, "Are you not satisfied with what you did? How much more do you want to torture us? What more is left?" and started weeping. This was the reaction of the majority of the camp residents.

The 3 women from Sakhi along with 3 from Saath formed 3 teams, each team comprising of a Saath and a Sakhi member.

Despite this strong resistance and threats that they would be harmed if they visited the camp, they did not falter in their mission. They continued going to the camps and talked to the people (women and children), explaining to them that all Hindus were not bad and they genuinely wanted to help.

Simultaneously, members of another Saath facilitated CBO, Ekta Yuvak Mandal joined hands with Saath and Sakhi in their endeavor. Old links were also revived with the people at Sankalitnagar. The founder of Saath had been associated with Sankalitnagar since 1983, while working for St. Xavier's Social Service Society (SXSSS). The young men with whom the founder had worked with, had now become leaders of their community who commanded respect. They had now initiated the formation of a CBO, Sankalp Mitra Mandal, and were committed to work for their area. Discussions with these leaders were fruitful and they were ready to work together with Saath.

Work in the Relief Phase

Out of the 101 camps in various parts of the city, Saath took up rehabilitation responsibility in the camps of Jamalpur, Behrampura, Ramol, Saraspur, and Juhapura.

The Saath-CBO teams witnessed the deplorable condition of the camp inmates. There was lack of sanitation, hygiene, food, clothes and other basic necessities. Thus, when Citizen's Initiative divided its work into various sectors, we took up the responsibility of procurement and dispatch of food materials. All traders and wholesale markets were closed. Procurement had to be done from retailers, making the job quite challenging. Availing of transportation was another challenge. Through a great deal of perseverance, 113 tonnes of food items were distributed to 37,050 people in 29 camps within the first 6 days.

The Saath team visited and interacted with the inmates of the camps through the day, in an effort to ease their sorrow. They played with the children. It was shocking to see that the games the children played were based on themes of violence and destruction. As in any disaster situation, the most affected and vulnerable are the women and children. It was essential to initiate constructive activities with the women and children to divert their attention. The gruesome incidents during the riots had left the children with a fear psychology. This aspect had to be taken into account and solutions found.

The issue was discussed with the camp inmates and the idea of starting activities for the children was put forward. The women especially welcomed it. Thus, under the Save the Children Programme, activities were initiated with children in the age group of 3–12 years. Playing kits and nutrition support was provided. Women from the camps were identified and trained as teachers by the CBOs. Saath worked with young girls too, through vocational training classes in hand embroidery, mehendi, patchwork, health and hygiene.

Saath also participated in the Aman Samuday programme supported by Action Aid. This programme aimed at establishing peace and trust between the two communities. Members from both the communities were trained as Aman Pathiks. They collectively helped in resolving problems of the camp inmates.

The task taken up was not easy for the Saath and CBO teams. At every step, they faced resistance from their family and neighbours. Some of women from the CBOs did not inform their families about their daily work at the camps. Eventually, when the fact emerged, there was a great deal of turmoil in each household. But, they were determined to carry on with their work. Not all were able to convince their families; however, their steely determination kept them going.

The community, as a whole, was against women working in the camps. One incident that comes to mind – one morning when Sakhi member, Devuben, was going to the camp, the people of her area stopped her and questioned her as to where she was going. To avoid arguments Devuben gave them an elusive answer. The people immediately told her they were aware that she was going to the camp every day. They said," What are you getting out of this? Aren't there enough people of your community to help that you need to go to the camps to help them? Do we not have needs? Muslims are not trustworthy. Are you not scared? What do you want to prove? You should not go to the camps to help them." Devuben explained to them that there was no cause for such hostility. She spoke to them about how the people were suffering in the camps; how they had lost everything in the horrifying communal violence. She also told them that the camp inmates would not harm her in any way; they were in a helpless situation and in need of support and assistance. She also made it clear that nothing would break her conviction to work towards peace. The people finally made way for her to go to the camp.

Work in the camps continued for months. Slowly, people went back to the chawls in different areas. Many moved away from their old areas and shifted to areas, which they felt were relatively safer for them.

Steadfast determination, as well as the various initiatives taken up in the camps by Sakhi, Ekta, Sankalp and Saath, not only helped alleviate the pain to some extent, but also initiated a thread of trust.

When the riot victims returned to their homes, we sensed the need to continue working with them as there was still a lot to be done for the people to get back to a peaceful, normal life. Organisations like CARE India and AIF supported us in the endeavour. Eventually, Saath worked in the following areas:

Areas	No. of Households(H/Hs)	Population
Sankalitnagar, Juhapura	523	3557
Chawls in Behrampura	844	2913
Chawls in Saraspur	340	2048
Santoshnagar	401	1965
Total	2108	10483

Work in the Rehabilitation Phase

Apart from loss of lives, the other two major losses were shelter and livelihood. An alarming fallout of the riots has also been the gradual ghettoisation of the city into Hindu and non-Hindu areas.

Shelter Support

Juhapura and the chawls in Behrampura and Saraspur posed a different kind of problem with regard to shelter. People from various parts of the city came to the Juhapura camp and eventually decided to settle there because it was considered to be one of the safest places in the city for Muslims. It is the largest Muslim ghetto. Thus, while Juhapura was not directly affected, it became a safe haven for innumerable riot affected families. On the other hand, chawls in Behrampura and Saraspur had been looted and burnt during the riots. The residents had vacated the slum and were staying in camps.

The Islamic Relief Committee (IRC) had committed itself to rebuild houses of the riot-affected families. Saath decided not to interfere in IRC's work. However, at Juhapura, due to the influx of refugees and their desire to settle there, rents shot up. Moreover, the landlords made a deposit mandatory. Hence, Saath assisted people in this area. Deposit money was provided to the landlords with an agreement with the residents that they would repay it to Saath within eleven months. Fifteen families were given assistance, ranging from Rs. 5,000 to Rs. 15,000.

Livelihood Support

Micro level surveys were conducted by Saath in the areas where it had decided to work.

A detailed analysis of the number of people affected, the extent and type of loss and, livelihood options, was carried out. It was decided, as a matter of policy, not to give grants, but provide interest free loans, which the beneficiaries had to repay in easy installments. A Community Fund was formed from the recovery thus made, which was to be used for future development of the area. The other benefit of this system was, it discouraged people from asking for more than their requirement, which helped kick start their lives once again.

Loan committees were formed for each area. The committees were responsible for identification and verification of the beneficiaries, signing contracts, and disbursing loans in their area.

The recipients of the loans started paying their installments. The task of recovery of the loans has been a daunting one for the coordinators. Several meetings were held to reiterate the purpose of the loan, after which people were convinced and repaid their loans.

Integrated Development Initiatives

Saath's reconciliation efforts have been through development initiatives. These initiatives made peace building an intangible yet definite part of the process.

Saath decided to continue its work on a long-term basis in the area of health, education, needs of women and children. The lack of proper savings as well as their dependence on moneylenders was another reason for their vulnerability. CBO members became the first line facilitators.

Thus, the second phase of Saath's intervention was its existing ISDP model, with variations to suit specific needs. The daunting task before the team was identifying local people, and training them in the fields of health and education. Women familiar with the area, interested in improvement and development, were selected as local workers. Sakhi members overcame the continued resistance from the locals through regular home visits, street meetings, etc.

In all the areas, activities were initiated depending on the demand, need, and feasibility.

Community Health Programme (CHP)

Saath's CHP is a holistic intervention aimed at providing preventive health care. It also attempts to deal with slum-specific health concerns.

Focus group discussions revealed that health problems like hypertension, migraine, insomnia, and other diseases, manifestations of the tension and anxiety, was on the rise. There were cases of children suffering from nightmares about the riots, enuresis, etc., which revealed the intensity of the impact of the rioting and violence on the children.

Initially, discussions with locals as well as field visits were carried out to identify the slum specific health concerns. For example, it was revealed that in Juhapura/Sankalitnagar prevalence of TB was high. Awareness regarding health and hygiene as well as mother and child-care was abysmal.

Immunization and growth monitoring programmes were introduced in January 2003 at Juhapura and continued at regular intervals.

At Behrampura and Saraspur, referral programmes were run for TB and immunisations. At Santoshnagar, a comprehensive health programme was started in July 2004. Gynecological camps were held twice a month.

A DOT centre was initiated in January 2004 in Juhapura, to identify and treat TB patients after the microscopic tests were carried out at Pravinnagar-Guptanagar.

In April 2003, a dispensary was set up in Juhapura. Subsequently, the health workers were capable of carrying out health activities independently in their area. They attended regular ongoing trainings. Home visits, street meetings, running of the DOT centre, dispensary and all other health related activities are being efficiently taken care of by the local health workers.

The residents benefitted not only by having health facilities in their vicinity, but also at affordable rates. Home visits were carried out; Hindu workers visited Muslim households and vice versa. What seemed like an impossibility, actually resulted in increased interaction between the two communities, with prejudices on the way out.

Non Formal Education Programme (NFEP)

NFEP aimed at decreasing the dropout rate of slum children in schools by helping them to cope with their existing school curriculum, inculcating in them a willingness to learn, and encouraging female education.

Two Balghars (Pre-Schools) were opened. However, there was reluctance on the part of the parents in sending their little ones to the Balghars. They were apprehensive about the programme. However, the devotion of the teachers brought about a change in the attitude of the parents. Gradually, the parents were won over as they saw change, for the better, in their kids. There are 15 Balghars running, including the ones in the riot affected areas.

Teachers from Sakhi trained the local women in the Montessori method of teaching. Further, special trainings were also provided. The Sakhi coordinators ensured proper planning, and monitoring of the activities. The women became confident and were able to conduct the classes themselves. As Sanjidabanu, a teacher in Balghar (pre school) said, "When I attended the workshop organized at AMA, it was the first time that I had come out of the closed confines of my home. I realised there was so much to know and learn in life."

Picnics and outings, an integral part of our non-formal education programme have also helped tremendously in establishing interest, credibility, trust and rapport. When the first picnic was held, parents were reluctant to send their children, especially girls, away from their homes. But today, there is demand for such picnics.

The mothers felt safe sending their children to preschools in their locality. The curriculum helped the children mentally get over the effects of the riots

Residents in other wards also demanded opening of more Balghars in their wards in Juhapura. This was a very positive impact, as our long-term plan was to work in all the wards, covering the entire Sankalitnagar.

Supplementary classes had also been introduced to provide additional support to school going children. There were 3 supplementary classes, one each in Juhapura, Behrampura, and Saraspur.

Quote of one of the pre-school teachers, "As a teacher, I had to go to many Hindu dominated chawls, an act which I would have never dreamt of doing in my life. It was an overwhelming experience – not only did I meet the parents in those chawls, but I could not return home without drinking seven to eight cups of tea, each home wanting to welcome me."

Four activity centres were started, one in each cluster. One of the aims was that it served as a platform for interaction between the communities. A committee was formed to oversee the setting up – books and games were selected through the committee. These centres catered to the residents, all age groups. Its functions manifold. It primarily served as a Library with newspapers, books and magazines for all. Indoor games, like chess, carom, etc. were an added attraction for the children. Further, there was a special area demarcated for children studying in secondary and higher secondary to be able to study undisturbed whenever they chose to. Competitions and sports events were conducted at the centre at regular intervals.

Livelihood Programs

The livelihood sector aimed at increasing the earning capacity of the slum residents by enhancing their skill base and making critical interventions in the employment market. The sector included skill imparting activities, income generation initiatives and formation of savings and credit groups.

Skill Imparting Programme

This activity aimed at enhancing the existing skills and imparting new skills to enable them to increase their income levels.

There had been a very positive response to the tailoring classes that had started out as a rehabilitation activity in the camps. After the camps were disbanded, we decided to continue the activity with a modified curriculum, as a six-month certificate course. The curriculum was carefully designed to suit the socio cultural needs of the Muslim community. Members of SMM identified and trained women from the respective areas to work as instructors.

The classes have become an important forum for young girls of different communities to interact and discuss various issues, which they cannot otherwise talk about in their conservative homes, including issues pertaining to adolescence.

These classes have increased interaction between the two communities at every level – teachers, students, and parents.

At Juhapura, the tailoring class started functioning from January 2003, but in the other areas, it started from April 2003.

Income Generation Programme

The programme aimed at identifying new areas of income generation with stress on critical intervention in the service sector. The programme was initiated in Juhapura in November 2003 when the young women procured an order for providing computer covers, dresses, etc.

In January 2004, there was a tie up with AWAG, who run an export centre for tailored and hand embroidered items. Women from the Saraspur centre were trained by professionals in sewing, tailoring, and hand embroidery to be able to produce good export quality items. Some women gave up on the tedious training and had to be persuaded about the requirement of quality work. 27 women were employed for hand embroidery and 5 were given specialized training.

Demands by the youth for new avenues in livelihoods were being explored.

Savings and Credit Programme

This programme helped reduce vulnerability of the poor by increasing their savings base as well as facilitating their access to institutional credit.

The SCP received tremendous response in the areas it was implemented in. The initial fears of being duped by the organisation of their hard earned money faded quickly. The credibility established by Saath in other programmes played an important role in alleviating their fear. But more importantly, the responsibility of running the programme was given to people from within the community. They were trained to handle the programme. When the residents feared losing their savings, the coordinators assured them the safety of their savings was their responsibility.

In Juhapura, where the activity originated, the number of members saving was more compared to the other areas. A children's saving scheme was also initiated. At present, the amounts saved by the members are collected by local workers and deposited in the two established societies of Sakhi and Ekta, (to be formalized and registered in the near future).

Physical Infrastructure and Environment Improvement Sector

Sankalp Mitra Mandal at Juhapura/Sankalitnagar took the initiative of collaborating with AEC (Ahmedabad Electricity Company) for provision of electric metered connections to the residents of the area. 875 connections were provided. Initially, the residents were not confident the money they were depositing (Rs. 5000 for an electric connection) was in safe hands. The coordinators persuaded them that metered connections were safe and legal.

Subsequently, the whole situation changed. The coordinators reported that the residents were so enthusiastic that they were even willing to go to the AEC office on their own to complete the necessary formalities. Worth mentioning here – most of them have also learnt the procedures!

Discussions for electrification of the other areas were initiated with the AEC.

Community Participation and Mobilisation

The crux of all development interventions is capacity building of the people. For Saath, the process of training local workers in various sectors to the extent that they were capable of handling various activities and programmes on their own was a very enriching learning experience. Right from the beginning Saath had proposed that members of SMM and EYM be torchbearers in the Muslim areas where work would be carried out.

Interaction between Hindu and Muslim workers helped in lessening the prejudices that had existed. There was a definite change in approach among the coordinators.

This, in turn, percolated to the people of the area. There has been a positive change. Through parents' meetings, street meetings, sports events and various other programmes, people of both communities shared the forum.

Successes of Saath and the CBO's development processes had initiated interest in communities in other areas. They were eager to replicate the development initiatives of Sakhi Mahila Mandal in their areas.

Sankalp Mitra Mandal was registered in May 2002. In the other areas, the process of registration of CBOs was initiated.

Involving the youth in development processes was the need of the hour. In February 2003, efforts were made to reach out to them through sports. Volleyball was the game that engaged their interests initially. 20 young men formed their first team. Subsequently, a path-breaking event of playing matches against other teams was organised. 13 teams participated. This was true inter-community unity! The Hindu teams had been of the opinion that Juhapura was a "restricted zone" as far as they were concerned and never dreamt that they would be playing matches there – taboos were being edged out.

Sankalp's independent association with other CBOs, NGOs, and government bodies increased manifold. It has independently partnered with Saath and Sakhi in various programmes and meetings.

Sankalp established good linkages with the gram panchayats of Maktanpura and Vejalpura and coordinates with them for various programmes. It assisted in the mapping of houses that did not come under the Maktanpura gram panchayat. This helped the house owners get individual water connections, streetlights and other amenities.

Understanding Processes

When Saath initiated its intervention activities, it had not consciously decided any particular course of action. It responded to the need of the hour, after deliberations within and with the affected, as to the best course of action.

Looking back at the steps taken by Saath, it is clear that they were based on the demands of the situation. The environment during the riots with its changing dynamics was constantly closely monitored and became the foundation for positive action.

The communal riots led to a lot of soul searching at Saath. Saath's premise of development work has been empowerment of the poor and vulnerable. The history of communal violence in Ahmedabad, especially the present one, confirmed the belief that until issues of conflict resolution and peace building were addressed, any effort at development would be in vain. Thus, it was essential to work proactively in this direction.

Relief Phase

Saath initially responded to the situation by joining Citizens' Initiative and undertaking immediate relief measures. In order not to duplicate relief measures and simultaneously ensure that people were not deprived of any services, Saath undertook those activities that had not been taken up by other organisation.

Livelihood Support

A crucial decision that was taken by Saath during this time was that it would not disburse grants, it would provide interest-free loans which the beneficiaries could repay in easy installments. It was not an easy task to convince the beneficiaries or the funding organisations about the relevance of the approach, especially at a time when there were several organisations that were willing to give grants to the affected. However, Saath stood firm. The repayment money thus collected went back to the area and their associations for developmental activities. The disbursement of loans, and not grants, also ensured that people were discouraged from asking for more than their requirement. This course of action prevented dependency; it pointed them in the direction of getting back to their normal lives.

The focus group discussions, held in different riot affected areas, revealed that the worst blow was to livelihood and employment opportunities. Destruction of equipment, lack of investment avenues, relocation (far from their original dwelling place), and communal insecurity all contributed to the loss of earning a livelihood. At workplaces, some were not taken back due to communal differences, but for most, the factories had been badly affected, and did not have enough resources anymore to employ people on a large scale.

Integrated Development Initiatives

Saath's reconciliation efforts were made through development initiatives. Saath understood disasters affect the most vulnerable lower economic strata of the society. Empowerment along with development is the way for them to get out of their misery. Saath has realized and firmly believes that forcible peace initiatives cannot be successful. Peace building should be an intangible, yet definite part of the process. For these reasons, continued developmental work in these areas on a long-term basis was the answer.

Saath had experienced that issues to be tackled were best understood by the people who were going through the problems. In this case, slum dwellers. The CBOs experience in the area made them the best people to able to address the problems in a way that worked for them. The CBOs (which mainly comprised of Hindus) became the torchbearers in the riot-affected area.

Working primarily with the locals was the most beneficial way of development with the least obstacles and resistance. To further the process of peace and harmony between the communities, we made sure there was a fair representation of both communities in the programmes. In the Balghar programme, the two teachers for each Balghar were one Hindu and one Muslim.

Lack of livelihoods and livelihood options led to infighting, children giving up their education to help with the family income, and last, but not the least of the problems was, some of the youth whiled away their time and got into bad habits like drinking and gambling. As was to be expected, there were demands for more livelihood options. The positive outcome however, was that women joined the work force to help make ends meet.

Reconciliation and Community Development

The role of Saath, in the eyes of the riot affected subtly changed over time from being perceived as intruders, to the people they turned to, for help to address all their needs/issues. A bond of trust had been firmly established.

With resistance from every quarter, family and community, as well as with their own apprehensions, the Saath-Sakhi-Ekta-Sankalp team members, however, were steadfast in their mission of peace and development in the riot torn areas.

Juhapura, is considered one of the safest places for the Muslims in the city, in fact, the largest Muslim ghetto in the city. During the riots, Juhapura was home to a very large section of Muslims.

The camp inmates constantly berated the members of Sankalp Mitra Mandal (Muslims) for associating with a "Hindu Organisation" (Saath). Most of Saath members were Hindus and had not been proactive with the Muslim community prior to the riots.

This outlook expressed by them was food for thought. It led to introspection as an organisation, as well as individually. Over time, with regular interaction, a good rapport, mutual trust, and respect prevailed.

Working on development activities with the locals in the riot affected areas has been a positive learning for Saath. Many of the taboos and myths were slowly done away with and interaction between the communities took over. The unimaginable was taking place!

Community Based Organisations (CBOs) have been vital in making breakthroughs in the intercommunity harmony. They have been the driving force at all stages of development and peace.

An area wise review of the various processes would help gain better insights and help analyze the processes in greater detail, at a micro level.

Sankalitnagar/Juhapura

As already mentioned, Juhapura, a Muslim dominated area, was not directly affected by the riots. It, however, became a comparatively safe haven for a large number of riot victims.

Development in Sankalitnagar, in a way, had been faster than the other riot affected areas that Saath worked in, as Sankalp Mitra Mandal was already an established and experienced CBO working with the people of Juhapura.

The focus group discussions held in the area revealed that there was satisfaction with the development work carried out in the fields of health of women and children, education – especially for the girl child, as well as with regard to the physical infrastructures. Issues related to mental health – insomnia, headaches, blood pressure, etc., which had been rampant immediately after the riots, were slowly easing off. Livelihood opportunities, however, continued to be a growing requirement for the youth and for the women.

Other wards were reaching out to Saath for similar development activities in their areas.

Impact in the Affected Areas

Behrampura though affected by the riots, did not suffer major losses. No lives were lost. Education standards and literacy levels were comparatively high here, just 5.6% illiterate. They were comfortable accessing health services.

It was difficult for them to overcome the inhuman acts of violence that took place between the two communities. They were left frustrated, sad, angry, helpless and vengeful. Fear and mistrust peaked immediately after the riots.

With time, things slowly got better. The residents attributed their attitudinal change to the significant role Saath played post riots. The various programmes initiated by Saath paved the way for development in the area, especially with regard to women and children. Here too, the refrain was – creation of more livelihood options is the need of the hour.

Saraspur

The residents of Saraspur shared their horrific experiences during the riots. Houses were burnt and looted – incidents that they had never imagined, could happen in their area. The women reminisced their happy celebrations of all festivals and ceremonies together, before the riots. But post riots, they were no longer a part of each other's sorrows and happiness.

Gradually, attitudes changed. Consistent efforts by Saath, CBO members and local workers helped in this process of change and breaking of barriers. Trust in the Saath team has grown manifold.

However, for some there is the fear that the peace would be short-lived. Though the festivals are celebrated together, it has lost its total sincerity; they are a little wary of each other.

The CBO members have decided to play a crucial role by taking full responsibility for the development and peace on a long-term basis.

Santoshnagar

At Santoshnagar literacy levels and health standards were low.

The focus group discussions and interaction with people clearly brought home the fact that bringing about reconciliation would take time. There was very limited interaction between Hindus and Muslims. Further, the Muslims were apprehensive about going to the Hindu areas. Mistrust and suspicion weighed heavily on them.

Many children had dropped out of school/college for a variety of reasons – lack of economic resources, and fear of going into the city. A number of them had taken to drinking and gambling. Once again, their foremost requirement was livelihood and employment opportunities.

The CBO members have an enormous responsibility of working towards the development of the area, and, more significantly, working towards bringing about an attitudinal change in the people.

An Overview of Learnings at Saath

Today, when we at Saath, look back, we know that there's still a long way to go to the goal we set ourselves. Glimpses of successful milestones are paving the way to our goal. We have been able to stand firm in our ideologies and our approaches. Regular interactions with those we work with, and for, have helped tremendously in meeting their development needs. Their appreciation, acceptance, and trust in our efforts have been the driving force; are a confirmation that we are on the right path.

Self-empowerment of the people has been the solid foundation on which all the development activities have flourished. The CBOs are a case in point. All members are from the communities working together towards betterment through their own initiatives, networking with government and other organisations.

Motivating the youth to positive paths of development has been rewarding. Their zeal and enthusiasm has once again confirmed that we are moving in the right direction. Another encouraging and reassuring factor has been women's participation in the process.

Being able to encourage and facilitate Hindu-Muslim interaction at the CBO and the local levels has given us the hope that this bond between the two communities can be strengthened for a better and prosperous future of the people and society at large.

For the future, emphasis on participation of the youth, as well as women, across all religions, for development initiatives are topmost priority.

Another successful and positive development initiative is the formation and working of CBOs. We look to the empowerment of the CBOs to the extent that they have an important political say in the issues relating to development and peace in their respective areas. Building a secular federation of CBOs working in tandem towards sustainable development in social harmony is the need of the hour.

The process has, undoubtedly, been slow and time consuming so far. Development through self-empowerment is a slow but effective process that ensures sustainable, long-term gain. The results, however, are not always tangible, but the process itself is an enriching learning experience. At this stage, people have started understanding the concept of sustainable development. Building on the trust and acceptance, further processes will be easier and faster. Soon, Saath's vision will be a reality!

Case Studies

Activities with children in the camp: Devuben and the little girl's story

When Devuben, a Hindu, member of Sakhi, used to visit the riot-affected camps, a camp inmate, a little girl, would start crying the moment she saw her. No matter how much Devuben tried to talk to her or play with her, she could not be coaxed into interacting with her, worse still, she would cry. After considerable deliberations, another member of Sakhi suggested that Devuben remove her "bindi," (a traditional red dot, a symbol the Hindu women wear on their foreheads) before approaching the girl. When Devuben did this, the little girl actually stopped crying.

Devuben discussed the matter with the girl's mother. She explained that her daughter had seen a group of Hindu men and women with the red bindis on their foreheads attacking Muslim people and torturing them. She associated anyone with a bindi to those horrific images, and was terrified of them.

This incident is just one example of how the violence and barbarism had been sharply imprinted on the minds of children.

Sewing classes in a camp: Sairaben Pathan's story

Thirty six year old Sairaben Pathan, who has studied up to the higher secondary, lives in Sankalitnagar, Juhapura. Her husband's hotel business at Ognej had been destroyed in the riots of 1993, compelling them to shift to Juhapura. Her husband was unable to recover from the heavy losses incurred. Saira took tailoring orders at home to make ends meet.

At the relief camp, she came in contact with members of SMM who persuaded her to become an instructor in the sewing class being run at the camp. After the camps were closed down, they decided to continue the classes. However, it was to be a formal training, which meant a proper syllabus and regular classes. Trainees would pay to learn.

The sewing class was a success; for Sairaben, the respect that she was shown by her community was very vital. Her financial contribution to the family increased. Most importantly, she had found the space to express her creativity. She decided to pursue further studies, and enrolled for a Bachelors Degree in Arts.

Aman Samuday: Zoraben from Santoshnagar

Zoraben, 36 years old, from Santoshnagar, firmly believed that their area would never be affected by riots or violence. She was shocked when violence broke out in her area too and she had to take refuge in a nearby Masjid. Her house was looted and burnt. Her children cried and asked for their toys left in the house. Zoraben was very frustrated and sad.

At the camp, she met the Saath/CBO team and had lots of discussions with them. This helped her vent her sorrow and anger. When the Aman Samuday programme was initiated, despite resistance from some of the camp inmates, she became an Aman Pathik. In her words, "I had suffered so much, lost so much. I didn't want this to happen; I have had the opportunity to vent my feelings, and now I want to help others who may have suffered much more than I have."

Livelihood support: Abdul Kadar Abdul Karim's story

Abdul Kadar Abdul Karim, 40 years old, has been residing in Jethalal ni chali, Behrampura, for past 30 years. He has a family of ten to support. He earned his living by repairing and making new handcarts.

During the communal riots, his entire house was looted. His business too was completely ruined. He had no resources to restart his business. This helplessness had a terrible effect on his health.

Abdulbhai was assisted with a livelihood support loan of Rs. 10,000. Other family members were also helped in different occupations. Today, Abdulbhai earns about Rs. 2000 a month. Abdulbhai said, "Not being able to support my family, made me terribly depressed. I have started earning today, and that gives me hope for tomorrow."

Mohamadbhai's story & Tahira apa's story

Mohammedbhai in H Ward, House No.205, Juhapura, was a much harassed T.B. patient doing the rounds of various hospitals – in Astodia the V.S. Hospital, Arogya Kendra in Vejalpur, etc. but, in vain. He was identified during the Saath health workers home visits to identify potential patients. With regular monitoring and medication as well as nutritional support, he has been improving.

Tahira, 31 years old, from Juhapura, spent Rs. 90/- at a local dispensary for checking on her pregnancy, but was not satisfied with the check up. She visited our health centre. She has 3 daughters and did not want another child, scared that it would again be another daughter. A pregnancy test was conducted here; she was given a positive answer. Through constant counseling, the health workers convinced her not to abort the unborn child, as it would have been detrimental to her health. Moreover, she was also counseled to take precautions to avoid pregnancy in the future. She delivered her child and is happy at the personal care and guidance given to her.

Sakirbhai's, Hurabibi's and Hamidaben's story

Sakirbhai, a resident of Saraspur, lives in a joint family of eight members. He had a garage and earned a good living before the riots. During the riots, the garage was destroyed and he lost his source of livelihood. The other family members also could not help him much as they too were badly affected. Sakirbhai took a loan of Rs. 10,000 from the Saath loan committee. He restarted his garage business. The garage is doing very well. He is able to adequately augment his family's income. One of his nieces attends the Balghar run by Saath in the area. Sakirbhai has not only repaid his loan, but is also saving with the SHG.

At Behrampura in Jethalal ni chali, Hurabibi's husband, who was a pedal rickshaw puller, was killed during the riots. They were a large family of five daughters and one son. Three of the daughters were married. The other three children were studying in school. Hurabibi now had the responsibility of supporting the family. She started a tailoring business with a loan from Saath's loan committee. She is now self-sufficient and repays her loan regularly. She supports her family and is able to send her children to school. Moreover, she also saves with the SHG.

A young lady, Hamida, made her living taking tuitions at home. She was initiated into the savings & credit programme in Sanklitnagar. Initially, to develop trust in the community she made home visits to explain the benefits of savings, but the response was poor. Gradually, with the introduction of the preschool & skill imparting programmes, people began to trust in the system.

Mumtazbanu, who refused to listen to her and influenced her neighbours not to save with Hamida, later initiated her savings with Rs. 1100/-. The other ladies who were ashamed to approach Hamida because of their earlier stance, approached another local worker to start their savings account. Hamida is happy to announce that more than 12 members have joined the savings & credit project due to her individual efforts. Presently, people from the locality go directly to the Mandal office to open their respective accounts.

Mohanbabubhai's and Sakinaapa's story

Mohanbabubhai, was a Hindu resident of Suleiman Roja ni chali at Saraspur, a predominantly Muslim chawl. When he passed away due to a cardiac arrest, none of his relatives were able to make it in time for his funeral. His Muslim neighbours and friends made all arrangements for the funeral ceremony, as well as accompanied his body to the cremation site for the last rites. They felt that it was their duty to help their neighbour in need. There was a furore when the Hindus at the cremation site saw so many Muslims. However, when the situation was explained they co-operated and appreciated the gesture.

After the riots, Sakina apa and Fatima another CBO member at Behrampura harboured strong feelings of revenge and hatred against those Hindus who had looted and killed the Muslims during the riots. She had witnessed a lot of violence and had had a narrow escape when a bullet passed just inches away from her leg. After many interactive sessions with other CBO members and a psychiatrist, her attitude changed. As an active member of the CBO, she now works towards peace and harmony in the area.

DATA TABLE

Table1: Details of the various programmes, facilitated by Saath, in the riot affected areas in Ahmedabad

S.No	Programme/Activity	Name of the Area	No. of centres	Participants
1	Community Health Programme	Juhapura	1	
1a	Outpatient Dispensary			942
1b	Gynecological Checkups			220
1c	TB Programme			181
1d	Immunization Programme	Juhapura	1	322
		Santoshnagar	1	178
		Saraspur	1	209
1e	Growth Monitoring	Juhapura	1	223
		Santoshnagar	1	195
		Saraspur	1	133
2	Non Formal Education Programme			
2a	Preschool classes	Juhapura	2	80
		Santoshnagar	3	110
		Behrampura	7	190
		Saraspur	3	97
2b	Supplementary classes	Juhapura	1	37
		Behrampura	1	
		Saraspur	1	
2c	Activity Centre	Santoshnagar	1	24
		Behrampura	1	47
		Saraspur	1	35

Contd...

S.No	Programme/Activity	Name of the Area	No. of centres	Participants
3	Skill Upgradation	Juhapura	1	40
		Santoshnagar	1	17
		Behrampura	1	27
		Saraspur	1	30
4	Income Generation Programme	Saraspur		27
5	Savings and Credit	Juhapura	1	84
		Santoshnagar	1	106
		Behrampura	2	190
		Saraspur	1	162
6	Physical Upgradation	Juhapura		1600 households
7	Community Based Organisation	Juhapura	1	171 members
		Santoshnagar	1	Have been recently registered in June 2004.
		Behrampura	1	
		Saraspur	1	

Annexure 3

Review of Saath's Urban Initiatives by Shrawan Kumar Acharya, PhD in September 2007

Acknowledgements

The present review was undertaken in a very short time. It would not have been possible to complete the study without the support of many individuals and institutions who are associated in one way or the other with the development work of SAATH. Special thanks to, the community workers, project coordinators, other staff members of SAATH and Mr Gagan Sethi the member of the Board of Trustees. Rajendrabhai, for willing to listen to the criticisms and for providing all the information and logistics to conduct the study! Kaushik Mehta, Subendhu, Priti Shah and Jyoti Solanki were good in providing a critical insight and an alternative perspective. The common people in the slums for their collective memories and uncommon wisdom!

Saath's Urban Initatives

Introduction

The report is a rapid review of SAATH's urban initiatives in Ahmedabad. The basic objective was to critically asses the initiatives and identify issues in the context of the organizational goal, values, capacities and future strategies. In doing so, an analytical framework was evolved to understand the historical trajectory of the organization, its identity as reflected in its development philosophy, legitimacy, accountability, community intervention and empowerment processes; its organizational structure, leadership qualities, organizational support systems, including staff profile and infrastructure; resources, policy intervention strategy, networking abilities, organizational learning and restructuring, and innovations. The methodology involved desk review of relevant documents – mostly annual reports, and participatory appraisal of the organization, projects and development processes. In-house appraisal involved detailed discussion with the CEO and other important staff at various levels, including the community workers. Other stakeholders directly or indirectly associated with SAATH's work were identified and interviewed, mainly the beneficiaries, government officials, representatives of the funding organizations, and academics. Projects were appraised through key informant and focused group discussion involving the beneficiaries and the project staff. Wherever possible information was triangulated and verified.

Development Philosophy

SAATH's approach towards the uplifting of the urban poor is positive, bold and contextual with respect to the changing economic situation in the city, country and the world. It is not based on the nostalgia of the welfare state. Unlike others who lament the role of the market in marginalizing the poor SAATH sees opportunity in the market, which can be harnessed in uplifting the poor. It believes in the fact that poor have strengths and capacity to face the market and survive in the market. What needs to be done is to "teach them fishing in the turbulent waters and not give them fish". The role of the civil society is to help them realize their strengths and guide them to stand on their own feet. Only such an approach provides opportunity to innovate and enhance the capacity and confidence of the poor. It is not the sympathy and pity that the poor want but realization of their strength by facing the challenges head on. SAATH's social innovations like the Urban Resource Centre and Umeed, fully substantiates this belief.

SAATH believes that:

- Market is good for the poor.
- Poor have the capacity and potential to take advantage of opportunities provided by the market.
- Advocacy is important but more than that service delivery is important while working with the poor. Advocacy has to be inbuilt into the service delivery strategy.
- It is important to work with the government and the politicians. They are too important an actor to be ignored given their legitimacy and institutional capacity. Civil society needs to leverage the strength of these institutions.
- Partnership, based on the core competencies of different actors is important to address the concern of the poor.
- Traditional development approach needs to be re-looked and made contextual to the requirement of rapidly growing city economies.
- Innovations and institution building is important.

Historical Profile

From its genesis in 1989, SAATH has consistently grown and matured from a small field organization to an important actor at the city and the State level. There is no doubt in indicating that in the urban sector, it is one of the most legitimate organizations recognized by the people, civil society institutions, private sector and the government. From 1989 to 1993, it was in its formative stage and spent most of its activities at the grassroots level consolidating its place and internalizing and shaping its values and mission objectives with the community. It was the entry phase. 1995 was a major landmark event. This was the year when SAATH started participating in the Slum Networking Project (SNP) project in Sanjaynagar and Guptanagar. This was its first exposure to the slum rehabilitation mission in partnership with other organizations. SAATH learnt a lot regarding the slum up-gradation work and benefits, dangers and risk in partnership projects. Despite the withdrawal of one of the partners from the project in Guptanagar SAATH persisted, continued to work in the area, and completed the project successfully. In the process, SAATH gained confidence of the people, the Ahmedabad Municipal Corporation (AMC) and learnt the art of implementing large-scale slum rehabilitation projects. At present SAATH is one of the few NGOs active in SNP. Similarly, SAATH took the post 2001-earthquake disaster relief and rehabilitation work as a challenge in its learning and community building process. The work was confined in the rural areas of Kutch District but the work provided valuable lessons to SAATH in disaster relief and rehabilitation. SAATH also actively participated in the post riots rehabilitation work in riot-affected areas of Juhapura, Saraspur and Jamalpur in 2002. It was SAATH's first experience in communal conflicts and reconciliation efforts. The work with the victims was instrumental in shaping and internalizing secular values and mission in its work. Historical trajectory indicates that SAATH has boldly taken up new challenges when circumstances have demanded, even if it was not its core strength, for the general good and the larger society. In the process it has enhanced its learning, organizational capacity and gained recognition and legitimacy from the larger society.

Legitimacy

SAATH is a legally registered organization as a Public Charitable trust under the Charities Commissioner, Ahmedabad. As per the commissioners requirement it has a constitution and a formal organizational structure with proper definition and elaboration of roles, responsibilities and power of different functionaries. It also has Foreign Currency Regulation Act (FCRA) certification to receive financial grants and assistance from external donor agencies.

Over the period of seventeen years, SAATH is recognized as one of the important urban NGO in the state of Gujarat. It actively participates in the urban development work of the Municipal Corporation and is actively engaged with the policy work of the Gujarat Urban Development Corporation (GUDC) and the Gujarat Urban Development Mission (GUDM) formed as a nodal agency under the JNNURM. SAATH is actively participating in the formulation of Gujarat Slum Policy and the Urban Poverty Alleviation initiative. Other civil society institutions like the CEPT, IIM, and SEWA actively work with SAATH in addressing the urban issues. SAATH's initiative is also getting recognition from international donor agencies like Oxfam, CARE and the American Indian foundation. Nevertheless, SAATH's legitimacy mainly derives from over 50,000 marginal people in the slums of Ahmedabad with whom the organizations interacts either directly or indirectly through its various community intervention processes. It is in recognition of these works SAATH was awarded the Anubhai Chimanlal Nagarikta Puraskar in 2004 by the Ahmedabad Management Association for improving the quality of life of slum dwellers irrespective of caste credit and religion. Besides, SAATH is also a member of the "Credibility Alliance" which ensures good governance of voluntary organization based on international norms, reflecting SAATH's quest for excellence and professionalism in its work.

Projects

SAATH started its work in 1989 with the Integrated Slum Development (ISD) concept in Guptanagar in Vasna. Since then the other innovative activities have been started. Some of the innovative projects are related to health and education, micro credit, livelihood, slum up-gradation and community empowerment through the formation of community based organization (CBOs) and the urban resource centers.

The health program emphasizes both the preventive and curative services by creating awareness and making services affordable and easily accessible. The objective of the education program intends to increase levels of education amongst the poor and make public education system effective and reachable to the poor. The health and education programs are integrated at the community level. The important components include the TB control Program, Reproductive and Child Health Program (RCH), Jeevandaan Maternal and Child Survival Program (JMCSP), Integrated Child Development Scheme (ICDS) and Balghars.

The RCH project has an innovative concept of "link workers" from the community, who reach out to nearly 12,000 households in Vasna and Paldi wards. Through 190 Aanganwadis under the ICDS, SAATH reaches out to nearly 14,250 beneficiaries. 3 Baalghars in Sankalitnagar provides education to 105 students. This program has become strong after the riots of 2002. The JMCSP reaches to nearly 2996 women and 1589 children. SAATH is also working with nearly 50 street kids in collaboration with other institutions like Indi Corps and Indian Institute of Management.

Micro finance program intends to facilitate savings and affordable credit for low-income population. The initiative was started in 1996 in Vasna. So far, four CBOs have been formed to address this issue, the Sakhi Credit Cooperative Society in Vasna and Saraspur, Ekta Credit Cooperative Society in Behrampura and Sankalp Mitra Mandal in Juhapura. Until July 2007, the total membership in these cooperative was 9,017 with a total savings of Rs. 1,07,60,459. 919 members had taken loan amounting Rs. 74,82,110. The loan amount ranges from

Rs. 500 to Rs. 30,000. It is very interesting to note that the loans taken for livelihood and housing purposes is substantial in number. The cooperatives are managed by the local people and the recovery rates are as high as 95 percent at an interest rate of 18 percent per annum. The traditional lenders charge as high as 10 percent interest rate per month. The savings and credit program has benefited the poor in accessing financial services especially in areas like Juhapura where there are no banking facilities.

The livelihood program is one of the most recent, relevant and challenging program. The main objective of the program is to provide skill-enhancing opportunities to the poor so that they can look for alternate livelihood options provided by the market and become self-reliant. The main components of the livelihood initiative are Home Managers, Umeed, Home Care, Housekeeping, Affordable Meal, Security Services and Drivers. Out of these the Home Managers program and the Umeed has been operational, other activities are at various stages of implementation and are yet to become fully functional.

The Home Managers program started in 2000 and so far, 165 women have been trained and placed successfully. The program is very successful in high and middle-income localities like Satellite, Judges Bungalows, Vastrapur, Drive in, Paldi, Sarkhej Highway and Bopal. The demand is more than the supply. The program trains women between the age of 20 to 46 from all low-income communities many of whom are from Dalit and Muslim families. SAATH does not entertain the caste or religion specific demand of the clients because they are exclusionary social processes. Home Managers earn Rs. 2000 for 4 hours of work and Rs. 5000 for 24 hours stay home arrangement. This is a high earning as compared to traditional maids. Moreover, the trained women are well respected and secure. The economic, socio-psychological change is tremendous. The self-esteem of women goes tremendous positive change. However, despite the success of the project getting sufficient number of women trainees is becoming difficult due to limited awareness and sometimes due to social prejudices. SAATH is trying to overcome this problem by sustained out reach and information dissemination strategy involving the URC, CBOs and grassroots workers.

Umeed started in 2005 based on Dr Reddy's livelihood enhancement model in Hyderabad. It was also financially supported by the firm for three years. Umeed is a youth employability training Program. Slum youth are trained for three months on various trades. So far SAATH has completed six training program. Over 1,200 candidates have been trained. The program charges Rs. 500 per candidate for three months. The program is now supported by the Gujarat Government and the Ahmedabad Municipal Corporation. SAATH has also involved other NGOs like Parivartan in the program. International agencies like CARE and American Indian Foundation are also supporting this program. There are four centers in Ahmedabad. The model is also being implemented in other cities of like Vadodara, Morbi, Mehsana, Nadiad, Bharuch and Patan. The program has scaled up from the city to state level. The response to the program has been tremendous and the demand very high amongst the youth. It is very encouraging to note that the number of women candidates is very high. In Wadaj center, the women candidates constitute nearly 60 percent of the trainees. Post training placement is as high as 90 percent. Most of the placement is in Ahmedabad based industries and service firms. The post training monthly earnings vary between Rs. 1500 to Rs. 7000. The program has innovative features like the involvement of the faculty in enrollment and placement for which they are provided incentives. The program has a built in mechanism to get feedback from the community through regular parents teachers meeting and post placement appraisal from the industries.

SAATH started its involvement in the slum up-gradation activity in 1995, with the Slum Networking or the 'Parivartan' project of the local authority at Sanjaynagar in collaboration with AMC and Sharda Trust of the Arvind Mills. So far, Parivartan has been successfully implemented in four slums and is nearing completion in two. For SAATH, SNP was the first experience in handling hard issues of infrastructure, and housing for which community mobilization and participation was important. Their role in community organization process was crucial for the success of the project. The first project was a learning process for SAATH in hard issues like infrastructure and housing provision. It was also an opportunity for them to operationalize the partnership model and evolve ethos and culture to accommodate the diverse demand of the partner organizations. The initial difficulties were gradually resolved by internalizing the learning from the project. Despite difficulties, SAATH continued to collaborate with AMC for SNP and today is one of the few, besides Mahila Housing Trust (MHT) and World Vision, reputable institutions involved in the project. SAATH has linked its other innovations in this project successfully and many of grassroots workers and participants in livelihood programs come from these slums. SNP experience has helped SAATH to scale up its activity and this has enhanced its visibility and recognition at the state, national and international level. The impact of SNP is spectacular in terms asset creation, livelihood opportunities, basic health and sanitation. Women have specially benefited from this program with the provision of facilities like water and sanitation and they have been able to spend more time in income generating activities.

The URC intends to be a change agent as an information collection and dissemination centre for the poor. In a globalized information age, lack of information is a major constraint in empowering the poor. This gap is being filled by the URC. The URC will assist in economic, social and political empowerment of the poor. They will be aware of their rights and duties. The work will also help advocacy activity of SAATH. Four URC has been planned, out of which the centre in Vasna is fully operational. The centers at Behrampura, Shahpur and Juhapura are yet to become functional. The URC will also facilitate interaction between the community and other stakeholders including the government and civil society institution. It intends to be a self-financed institution by charging for the services it provides to the community and other stakeholders. The information regarding various training program of Umeed will also be disseminated to the people at a nominal cost. SAATH has instituted proper institutional structure for the URC with a project coordinator heading the unit. Other members include the field coordinators, administrative and support staff. The field units are staffed by local community members. The concept has been supported by the AMC and many of its activities are being already channeled through the URC. For the AMC it is a very important support structure in its work with the poor. The functioning of the URC is overlooked by a strong committee consisting of lawyers, academicians, AMC chief engineer, local councilor, CBO leaders and local BJP leader. Such a structure not only gives legitimacy but will also make it effective. URC is also involved in the preparing micro plans and monitoring basic services through the Citizen Report card initiative. These activities help advocacy work. Researchers are also availing its services to collect information at the field level through traditional questionnaire survey and participatory methods. The URC also has the potential to play the role of watchdog as enshrined in the JnNURM. The URC has also lobbied on behalf of the Vasna citizens to bring adequate safe guard measures in the design and routing of the proposed BRTS route though the area. Complaints have also been registered with appropriate civic authorities regarding solid waste management and street lighting that has been redressed immediately. However, the URC concept is still in its formative stages and many local citizens are still contemplating its effectiveness and their involvement in its activities. Despite participation in meetings and planning activities, they are still not coming forward to take the leadership. Activity of such a nature requires volunteers who have to be from the community then only the information dissemination and advocacy role will become effective The URC can certainly be rated as one of the most important initiative of SAATH and proper planning and management is important to ensure its success.

The activity of URC is complemented by Community Video Unit (CVU) manned by local youth. The objective of the unit is to strengthen leadership, community mobilization, governance and advocacy through the power of media technology and participatory communication. The unit was initiated in partnership with Video Volunteers, a New York based organization, and Drishti Media. Short films, on community issues like communal harmony, health etc., and monthly News magazine is quite popular amongst the people. It is an effective strategy in sensitizing people. The unit also plays important role in disseminating information on livelihood programs like Umeed.

SAATH's relief, rehabilitation and development activity with the riots victims is an important initiative to address the issues of communal violence in the city of Ahmedabad, which has frequently witnessed such incidents in the past. This initiative involved CBOs from predominantly Hindu areas to start CBOs in predominantly Muslim area of Juhapura. The CBO, Sankalp Mitra Mandal, is still active and involved in community building activities supported by SAATH. Its activities have expanded to include savings and credits. A URC is also is also planned in that area which will be run by the CBO. Given the importance of the initiative CARE, has once again decided to the fund project.

Institutional Collaboration

The ISD program with which SAATH started its work in 1989 intends to integrate different activities and innovations at the grassroots level to minimize leakages. In the process, activities of different stakeholders are integrated depending on their capacities and core strength. SAATH designs and facilitates collaboration with other actors like the Government, NGOs or the CBOs. Such an approach has facilitated the participation of various stakeholders to work together at the local level. This has enhanced the effectiveness of the program in terms of both cost savings and outreach. The partnership model varies depending on the type of organization, their core strength resources and the projects. Therefore, the collaboration with Parivartan in Umeed is different from the collaboration with Akhand Jyot Foundation. However, the ultimate objective has been to ensure effective community empowerment and capacity building. In most cases, SAATH's partnership initiative facilitates resource convergence, reduces duplication and links various stakeholders. SAATH collaboration in the urban sector ranges from government agencies like GUDC, GUDM, AMC and various other departments like health education etc. Collaboration with international organizations has also been quite successful. Government and international collaboration is basically for leveraging financial and technical support. Resource centers like DHRISTI have collaborated in training community video workers. Collaborations with smaller grassroots organizations like the Ahmedabad Slum Dwellers Federation, Parivartan and Jyot Foundation has been for project implementation at the community level. SAATH in collaboration with the Indian Institute of Management and the Ahmedabad Municipal Corporation initiated a model Urban Health Centre in Vasna, which provides basic health services to the poor. This is one of the first innovations of its kind. In most of the cases, these collaborations have been successful. However, there are problems in forging alliances and institutionalizing long-term commitment. Working with the government means getting involved in and resolving bureaucratic structures and decision-making process. For NGOs like SAATH paper work and delays implies huge transaction cost. SAATH however has realized that many of these problems get resolved over time once the officials realize the importance of the work and understand the NGOs. The number of alliance SAATH has forged indicates that the organization has developed skill in building partnerships.

Engaging the Government Officials and the Politicians

SAATH also has strategic and balanced approach in relating with the Government agencies and politicians. They try to maintain a "healthy distance" and associate with them depending on the need and the situation. They again do not go by the common notion that all politicians are bad and government does not work! Politicians are legitimate representatives of the people and they have to be engaged in constructive dialogue. Similarly, the government institutions and officials have strengths, which need to be engaged constructively. Antagonizing the politicians and officials jeopardizes the initiatives. SAATH over time has cultivated and refined the art of negotiating and engaging with the politicians and government officials, which is reflected in number of successful collaborative programs that has enhanced the image and legitimacy of SAATH and bridged the gap between the people and other powerful actors of the society. Such an interaction has helped in changing the stereotyping and misperception of the "other" resulting in better dialogue between the development actors. The URC committee is a good example that intends to involve the officials and politicians in the activities of SAATH!

Community Engagement

SAATH has evolved a spatial strategy to reach the poor and the marginalized that is gradual and systematic. Largely this strategy is determined by the limited resources, financial and manpower, at the disposal of the organization but it allows the NGO to plan and phase its activity effectively. The strategy involves working with the various anganwadis, at present 190, in different wards of the city where poverty rates are high, in child and maternal health program. This is the entry strategy in the community. Child and maternal health being a major problem also ensures SAATH legitimacy and co-operation from the community. Only after the program has matured and SAATH has built considerable rapport with the community other programs under ISDP is linked with the anganwadi. This approach is being followed in expanding SNP work. Such a strategy also helps SAATH to plan and phasing out strategy.

SAATH's strategy to involve the community is based on the fact that community involvement succeeds only in those areas where the marginalized are more and where the need for such intervention is felt. Therefore, pre intervention strategy is to identify the marginal communities and their need. As indicated earlier, health, especially mother and child health that is the most important problem in slums, has been the most important entry strategy in the community. Such a strategy ensures participation, especially of the women. Many of these early participants are groomed by SAATH as community leaders, which give legitimacy to the program in the community and minimize conflict. Over time, CBOs are formed involving the people. Such a strategy has worked very effectively and CBOs are the strength of SAATH's success.

Community Empowerment

As indicated, most Programs at the community level are started after proper reconnaissance surveys to identify the needs, social, economic structure, political affiliations, conflicts and contestations. Attempt is always made to involve the community and only after ensuring their involvement, service delivery work is taken up. Community ownership of the projects is very important strategy for the empowerment process. This aspect is reflected in number of functioning CBOs in different slums like Vasna, Behrampura and Juhapura. The CBOs in Vasna are the strongest and are functioning effectively looking after various activities like the savings and credit, RCH and URC. The CBOs in Vasna are nearly 14 years old. The CBO manages all its day-to-day activities. Over time, they have become less dependent on SAATH for their routine activity. The CBOs from Vasna actively participated in rehabilitating riots victims in Juhapura. Four new CBOs were formed by older CBOs in riot-affected areas with minimum guidance from SAATH. The formation of CBO has not only linked two slums but has helped reconciliation initiatives between Hindus and Muslims of the area. Small community initiatives like this can help prevent future communal conflicts because a channel for communication has been opened between the communities.

Community Leadership

One of the most successful results of the SAATH initiative has been the creation of a cadre of strong and effective community workers especially in places like Vasna where the organization has been working for more than 10 years. In these places, the CBOs like Sakhi Mandal Vasna and Sankalp Mitra in Juhapura are strong and have played very important role in community mobilization. They are also very effective in disseminating information and actively participate in service delivery works like RCH and Savings and Credit programs. Local committees monitor the RCH Program and train the link workers. They are also instrumental in making the URC successful and effective. Moreover, some of these CBOs are also instrumental in the formation of new

CBOs in other places. The CBO in Juhapura after the riots was basically formed by the CBO from Vasna.

Changing Attitude

SAATH initiative has brought changes in the attitude of the local people. The strategy to involve people from different caste and communities in the CBOs, like micro finance or SNP has made them tolerant towards other caste and community, which was not there before. In some cases, SAATH engaged Dalit women as community health worker. They have been accepted by the community and are not discriminated. The home managers who work in high income upper caste families also come from such marginal groups. After the riots in 2002, SAATH started the relief centers in Juhapura which brought the Hindu and Muslim community together. In fact, the CBOs from Vasna were instrumental in starting the CBOs in predominantly Muslim areas of Juhapura. The strategy has not solved the communal issue but has certainly facilitated in bringing some members of the community together who can play a catalytic role in bridging the divide between the two communities.

The health campaigns amongst the women through the Reproductive and Child Health (RCH) Program and Balghars have led to increased awareness amongst the women regarding maternal and child health. In the initial stages SAATH workers had to go to individual house and deliver health services now people are more demanding they come to SAATH centers and seek health services.

The change in the attitude is also seen with respect to savings and credit behavior of the people. The number of people seeking the services under this category has increased substantially. The benefits are amply reflected in the number of members, savings and credit off takes. The credit has been used for various purposes but significantly for asset improvement like housing, especially in SNP slums and also for livelihood and income generation. What was significant to note was that the attitudinal change amongst the parents is also having positive impact on the children.

Innovation Adaptation and Scaling Up

One of the most important contributions of SAATH is in terms of experimenting with innovative concepts and scaling it up at the city and state level. This is amply reflected in programs like Umeed. SAATH internalized the Umeed concept from Hyderabad and experimented as well as adapted it to suit the local context. It built partnership with community, NGOs and the Government agencies without whose support scaling up and spatial expansion is not possible. The innovative URC concept has the same potential. The Ahmedabad Municipal Corporation has realized its potential and has already forged partnership with SAATH to implement its program. More URCs are being opened in various parts of Ahmedabad.

Service Provision and Advocacy

SAATH also believes that advocacy is important but for the poor access to basic services is important. The entry point strategies have to be by meeting the basic requirements of the poor. Advocacy strategies have to be built into the service delivery strategies. The poor want to see improvement in daily existence. A service delivery strategy legitimizes the existence of the organization in the initial phase after which advocacy strategies can be planned. Conscientization is important but improvement in the living standard is critical. This can happen only through innovative service delivery strategies, which SAATH has amply demonstrated through its participation in programs like the SNP or the innovation of concepts like URC, Community Video or the livelihood generation program Umeed.

Sustainable Financial Innovations of Projects

In order to ensure financial sustainability and also to reduce dependency of the community on free doles SAATH has a system of charging users fees from all the beneficiaries. This again is a reflection of its belief that poor should not be made dependent. This is important to bring attitudinal changes in the poor. Analysis of various program indicates that in SNP, the poor pay around Rupees 2000/household; in Umeed-Rupees 500 per youth per training; Home Managers training-Rupees 2500 to 3000 per person; pre-school charges per children Rupees 50 per month, and charges for the services in the Urban Resources Centre vary with the type of services availed. It is interesting to note that the concept of payment has been introduced even in the Pre School for children aged 3 to 5, which is free in many government run units. In the Positive Deviance/Hearth Nutrition Program, a mother spends Rs 4 per child per day in nutrition and health related activities for 15 days as a part of the special camp lasting 15 days. In the 15 days, the community member manages all activities with little support from SAATH. This not only builds peoples capacity to manage events but also enhances their awareness regarding health and nutrition and changes their cultural rigidities. In fact this experience has shown that poor people are more than willing to pay for quality services, moreover payments have also lead to greater involvement of the parents in the running of the schools, the nutrition quality has increased, pilfering of the food is absent and overall functioning of the schools have improved. The URC has also designed innovative way to meet some of its expenditure by charging for the services it provides to the clients. Rupees 50 from every Umeed participant go to the URC for the services it provides in terms of information and forms for the training program. The Vasna Sakhi Mandal meets 70 percent of its recurring expenses like salary, rent and electricity from the profit it makes through the credit activities. The Home Manager unit also meets substantial amount of its expenses by charging the trainees and also the client. Five percent of the monthly salary of the home managers is deducted every month as Mandal charges. The training expenses are also deducted from the monthly salary of the home managers. The subsidy component is little which makes the model successful.

Besides its project, SAATH has also taken initiative to persuaded formal agencies like Torrent Power Company to provide electricity to the slum households. The company found it a profitable venture and today it charges Rs. 2500 per connection, which was Rs. 9000 in the initial stages due to higher perceived risk by the company. The households are happy as they get the electricity. Payment is regular and for the company it has become a profitable venture to invest in services for the poor. The people have also realized the benefit of legal electricity, which is cheap and regular. They are not at the mercy of the intermediaries. Regular supply has also helped children's education and families are able to start income generating activities. SAATH has gradually withdrawn as an interface between the company and the households.

Withdrawal Strategy

SAATH intends to make community self-reliant. However, for SAATH, self-reliance does not mean complete withdrawal from its engagement with the community. It means taking up new roles and responsibilities, for the CBOs and SAATH, in the changed context. The new roles and responsibilities are discussed at the monitoring and review meetings involving the community. It has been observed that the intensity of SAATH's involvement reduces over time with increasing maturity of the CBOs. In mature CBOs, SAATH acts as an advisor and facilitator. It also helps them network and interact with new organizations. The CBOs take up the day-to-day responsibility of running and managing the institution. Financial dependence also decreases in some CBOs, like the Vasna Sakhi Mandal, as they are able to mobilize resources from their activities like savings and credit. In some cases, SAATH has withdrawn from community due to other reasons like conflict within the community.

Organizational Structure

The organization is hierarchically structured with clear roles and responsibility. The hierarchical structure however is not bureaucratic. It is flexible and decisions making often is informal and quick. Interestingly the weak second level management and relatively strong field structure of SAATH facilitates and strengthens bottom up representations better than many other NGOs in the city. The Board of Trustees is at the top, and provides overall policy and strategic guidance to the organization.

The CEO is in charge of the projects and overall administration. He is not directly engaged in the projects. His responsibility is more in terms of identifying and conceptualizing projects, liaising, building partnership with various organizations, and arranging for funds. The CEO is assisted by the senior most coordinator who overlooks all projects. Occasionally the senior coordinators act as a bridge between the lower staff and the CEO. This is however rare as most staff is easily accessible to the CEO.

The respective project coordinators have the overall management and implementation responsibility. The project coordinators are directly engaged in the field and are assisted by CBOs who is headed by community organizer and other field staff. Day to day field activities is the responsibility of the CBOs and community organizers. Occasionally, especially in case of project coordinators, there are instances of communication gap between them and the grassroots worker. Roles and responsibilities at each level are clearly defined though sometimes there may be overlap especially in areas where two programs are running simultaneously. This however has not created any conflict amongst the staff. In most cases such overlap are managed and resolved by the community workers or the project coordinators. The problem however, especially at the lower level, is related to multiple and unscheduled work which crops up frequently. As the activities of SAATH are increasing, staff has to take up work that is unplanned for the day. To some extent, good team effort has resolved such problems.

One of the problems that need to be addressed is the coordination between the project units and the administrative unit more especially the finance. Very often, the administrative unit fails to understand the difficulties of the field units. The adherence to bureaucratic norms disheartens the field operators. Sensitizing and orienting the administrative staff to field realities is important. Norms and rules need to check misuse but it should not constrain the work and de-motivate field staff. Rules need to confirm to standards but flexibility also has to be innovated.

The community empowerment strategies of SAATH to a large extent have been facilitated due to creation of proper organizational structure of the CBO with proper delegation of roles and responsibilities. In places like Vasna and Juhapura, the structure is manned by strong, confident and experienced women from the community.

Leadership

The founder director of the organization has evolved and matured over 20 years of grassroots experience with the marginalized of Ahmedabad. He has a good theoretical, contextual and practical understanding of the marginalization process. He is committed and passionate about the work. His ideas regarding the empowerment of the poor is contemporary and based on utilizing the positive processes of the market for the benefit of poor. This value has also been internalized by the lower staff. He has evolved a work culture that is flexible and allows scope for dissent and dialogue amongst the staff. He has the ability to take risk and innovate social concepts that are different from the traditional approaches, which are amply reflected in projects like Umeed, Urban Resource Centre (URC) etc. Innovation from the lower staff is encouraged. The Director is easily accessible to all staff members at different levels including the community workers. The vertical and horizontal expansion of SAATH's work however may be a limitation in the future. Increasing involvement of the director in policy and advocacy work will also limit his engagement at the grassroots level.

Staff

The number of staff working in the urban sector is around 100 out which maximum work in the field. Besides this, there are nearly 400 part time workers at the field level for various programs like the RCH. The livelihood program, Umeed, has additional 45 people. Overall, the staff profile both direct and indirect is quite big requiring proper management. The staff profile is inclusive representing the Dalits, women, Muslims along with other non-Dalit Hindu workers. Women constitute 91 percent of the staff and 73 percent is from the Dalit community and 12 percent represent the Muslims. Besides the permanent staffs, there are consultants and interns who are associated with different activities for short period. The interns are mostly students from India and abroad. Internship has enriched mutual learning and sharing.

Given the rapid expansion of work in the urban sector there is a need to enhance the human resource base in terms of both skills and number. At present, there is no permanent trained urban specialist, urban planner or urban community development worker. The present sector coordinators are experienced, educated and dedicated but formal professional qualification in the urban area is lacking. Many of them are also new to the organization retaining them for long term is important. Though consultants fill in the gap, it is not a permanent solution. Limited staff and expansion of work has overburdened the senior coordinator who not only has to coordinate different sectors and visit field but also has to takeover responsibilities of the Director in his absence, which incidentally has increased significantly in the recent years.

Staff emoluments are better in SAATH compared to other organization engaged in the development sector in the city. However, it is still less than other sector as a result retaining qualified professional for long is a concern. The strategy in SAATH to overcome this problem is to provide a work atmosphere that is informal, flexible and challenging. However, such a strategy does not guarantee staff retention. Incentives in the form opportunities to enhance learning through training exposure visits may be useful. Thought should also be given to replace uniform salary by performance based salary structure right across the organization including the field workers. This will reward the honest workers and motivate the non-workers. This strategy will fit into the market model that SAATH propagates. However, finances will have to be arranged to meet additional expenses if required.

Second Level Leadership

At present, second-level leadership is lacking and this gap may increase in the future unless serious attention is given. This may affect organizational efficiency. There is no second level person who can replace the CEO in higher-level meetings at the local or the sate level. The review indicates that the existing second level personal have good field experience and can handle the problems but their capacity to engage with other civil society institutions and the government at higher level is limited. Language and communication to engage and negotiate appears to be a constraint. Though there is system of weekly, monthly and annual consultative meetings, major decision are still centered with the CEO. Rapid expansion of SAATH's work in urban and rural areas also necessitates the importance of second rung leaders in the organization who can take over the responsibilities of the CEO in his absence. Grooming of competent second level leaders and delegating responsibilities needs to be paid attention. It is also important to highlight that increasing spatial expansion and sectoral growth of SAATH's activities may be distancing the founder members and the director from the grassroots people and staff. What corrective action needs to be taken will depend on the future work and organizational model that SAATH intends to follow.

Staff Capacity Building

Discussion with the staff revealed that SAATH provides them opportunity to build their capacity through in house and external training programs in various areas like PMES, research, communication, advocacy networking etc. This was true for the community workers also. Many of them have attained various training program in the city. However, it appears that the training programs have to better identified and organized. It needs to be in built in the annual plans so that other work plan is not affected.

Orientation

On-the-job orientation of the new staff exists but has to be made rigorous so that the candidates internalize the values of the institution and also come to know about the various activity of the organization. This will also help him/her to contextualize their role in the context of the larger organizational values. As the number of staff increases the need for proper orientation will be felt at all levels including the coordinators and field workers of the respective programs.

Internal Reporting and Decision Support System

SAATH has evolved a proper reporting and decision support system. Annual meeting is held to review the overall achievement and future directions. This involves all staff members, including the CBOs and community workers. Besides the annual meeting, two review meetings are held to monitor the work in each sector. Besides this, every sector and projects have monthly and weekly review meetings. The weekly meetings monitor the day to day functioning of the projects and resolves conflicts if any at the local level. The weekly project specific meetings are held at the field office but the monthly, mid term review and annual meeting is held at the head office of SAATH. It is also interesting to note that written reporting formats are in place for individuals, groups and sectors. All meetings are properly recorded and meetings at the lower level feed into the higher level. The annual reports are also prepared based on the compilation of these meetings. SAATH has also prepared a detailed project inventory including list of the beneficiaries and the staff at the grassroots level for internal monitoring and review. The CBOs are well structured and report formats are standardized across al the CBOs.

The overall long term direction and policy matters relating to the organization are taken up at the annual meetings and the Board of Trustees. The annual meeting identifies new activities, funding sources, fixes target and implementing strategies based on the feedback of the respective sectors. All continuing projects, achievements and constraints are also discussed. Based on the overall organizational plan and the review of sectors the plans are prepared. Each sector plan is thoroughly reviewed especially focusing on the constraints. The outcomes of the sectoral plans are then disseminated at the community level through the respective CBOs in their weekly meetings.

The reporting and documenting system, at the field level, are still manual and proper MIS system needs to be evolved immediately. Most of the SAATH field offices have computers and computerization is taking place. However, except the head office internet facility has not been installed. Installation of internet will increase connectivity and integrate activities of SAATH. Internet will greatly enhance information flows within the organization, between different sectors and between the different locations in the city. Decision-making will be fast and efficient. The advocacy role of the URC will also considerably increase.

Process documentation

Discussion with the staff and also the record indicates that the system of process documentation of the various activities is weak, which needs to be strengthened. The present structure and manpower does limit the continuous documentation of the various activities. Moreover, the multiple activities in which the organization is involved gives very little time for the existing staff to document the process. It may be advisable to strengthen the process documentation system. Manpower in the unit can be increased so that they can take up the process documentation activity in a continuous basis. Process documentation will be useful for orientation of the new staff and will also assist in training, organizational learning and restructuring if required in future.

Research

The research unit is still in its infancy. Research and publication, including action research, is weak despite having long, rich and diverse development experience. For this SAATH needs to strengthen the unit for research and documentation and hire competent staff. Though consultants are hired as and when required, it may be advisable to have in house capacity. System has to be created for proper information collection, storage, analysis and dissemination. Since SAATH is increasingly participating in policymaking process at the local and the State levels, in-house research and publication will be helpful. In this context, it is also important to highlight that the library needs to be improved.

Organizational Restructuring

Organizational character of SAATH has changed substantially over a period of time. It has moved from individual to more public institution with transparent practices. The power and function of the CEO has been considerably diluted through governance reforms. The Board of Trustee have been reformulated and made powerful and the CEO, called the Director before restructuring, is accountable to them. The current Board of Trustees represents the NGOs, Academics, Judiciary and other civil society institution. Each one of them is prominent in their respective field like management, finance, urban planning and sociology. The composition of management team also is inclusive in terms of gender, caste, communities and religion reflecting the work and ethos of the organization. The restructuring and empowerment of the management has provided new direction and legitimacy to the organization. The Board of Trustees also includes a senior SAATH member who has been working in SAATH for over 15 years. She started working in SAATH as a community worker. Her mobility up to the Board level is indicative of SAATH's basic values of empowering people and enhancing their capacities. She is in fact the grassroots voice in the Board and she still spends 75 percent of her time with the communities.

Program Restructuring

A conscious effort is made to reorient the Programs, its contents and the associated institutional structure to meet the changing demands and needs. The Business Development Plan prepared for the savings and credit program reflects one such initiative. The separate savings and credit groups in different areas are being merged to create a larger entity. It is also proposed to shift operations from the individual to joint liability groups to ensure better payment and savings habits. Merging different credit cooperatives will also help in standardizing rules and regulations and streamline accounting system across the societies. Separate auditing of all the cooperatives will not be there after the merger, which will save time and money for the organization. However, the most important advantage will be in terms of larger reach and larger loan amounts to the community. It will help scale up the activities that so far was small and community centric. Larger business possibility will also help the unit to become self-reliant. It will also help curtail other irregularities like double credits by the members.

Such restructuring has also been successfully attempted with the home managers program. It has been decentralized and a franchise model has been adopted to bring in accountability from the coordinators and is incentive based. The more home-managers the coordinators handle the more benefit they get. The system is also more accountable.

The changes have brought better management practice enhancing service standards and efficiency.

In order to make the livelihood initiatives self-financing the organization has promoted SAATH Livelihood Services Company under section 25 of the Companies Act. It will act as a not for profit company. The objective is to make the unit self sustainable under the independent management of the CBOs representing the urban poor. SAATH will only guide and facilitate governance. All other aspect like service delivery, day-to-day management will be with the CBOs. Such an approach will scale up the activities of CBOs and empower them in running small enterprise. This again reiterates SAATH's pro market philosophy.

Infrastructure

The support infrastructure is adequate in the main and all the field offices. Office space at the field is also adequate except in few places. In Vasna it is substantial allowing regular community meetings to be held, which appear to be difficult in Juhapura. The main office is owned but other places it is rented. In Behrampura SAATH uses the AMC space for its activities including Umeed training activities. Computerization has been achieved at the main and the field office. Most of the records and day-to-day activity is computerized. Interestingly the computers are manned by low-income person some trained under the Umeed initiative of SAATH. The computers at different locations, including the URC, has not been linked as the internet facility is absent except the main office.

Funding

The consolidation and expansion of work has increased the demand for finances for meeting the establishment and the project cost. In the year 2005–2006, the SAATH received Rs 41,171,370 from various sources and spent 36,696,307 on various heads. The utilization rate is quite high and satisfactory. Over the last 15 years, financial sources have diversified. In 2006, 50 percent of the resources were leveraged from the state 35 percent from the donor agencies and 10 percent from the beneficiaries. It clearly indicates that unlike many other NGOs, the share of national domestic funding is high but at the same time, the financial dependency of the communities on SAATH is very high. This is mainly because of the high service delivery component of SAATH with low contribution from the community. This is a cause of concern for SAATH especially because the organization has not been able to create a corpus fund that can act as buffer to sustain projects in case of cessation of external funding. This has been reflected in many projects. Lack of fund for TB and RCH is affecting the program adversely. Similarly, there is uncertainty regarding the funding of URC. Commitment of CARE to fund this initiative is until February 2008. The CARE funded Reconciliation Project with the riot victims was for sometime affected due to funding concern. Looking at the importance of the project CARE has once again decided to fund and revive the project.

External sources have also increased and diversified. The important organization funding SAATH activities are American India Foundation (AIF), Care (India), Cordaid, Counterpart International, India Friends Association (IFA), GIVE Foundation (Individual contribution) and IRC. SAATH has also made pro-active efforts to mobilize resources from external sources of which the linking up with the American Indian Foundation was a commendable effort. However, despite these there is still enough scope to plan and streamline the resource mobilization strategy.

Rapid Expansion of Urban Activities

In the last two years, the urban sector has come into the limelight due increasing emphasis on urban development by the Gujarat and the Central Government. Role of SAATH has increased considerably especially when the state is looking for partnership approach to resolve urban problems. Increased work has brought opportunities but new challenges as well. The sudden increase in urban activities have considerably stretched SAATH's work putting pressure on existing infrastructure, human and financial resources. SAATH is short of qualified human resources and unless this shortage is addressed, quality and efficiency may suffer. The renewed emphasis on urban also requires innovations in project delivery strategies which require thinking and research, which again is a cause of concern for SAATH.

Livelihood Training Concerns

The livelihood program is innovative and successful. In order to make it more effective some issues will have to be addressed. The current practice of training people in the age group 18 to 30 is excluding potential workers above 30 who can still be trained and made productive. Retraining model for them may have to be worked out. The other important issue relates to the time frame. It appears that 3 months of training is too less especially when the background of the students coming from marginal socio-economic group is weak.

Given the number of students, including many girls and faculty, the existing infrastructural facilities like classrooms and latrines appear to be inadequate. For students engaged in BPOs, the number of computers is also not sufficient for hands on experience. Curriculum of soma trades. Like ITES and BPO, lacks clarity. Relevance of some trades and courses offered to the students needs to be revised and up dated on a continuous basis. These issues are also true for other livelihood Programs. Information dissemination regarding the trades and training Programs needs to be strengthened.

Post Training Coordination Issues

Despite restructuring the problem of coordination, between the home manager and the client, still exists in the home managers Program. The coordinators mediate between the two. Every coordinator is linked to 40 home managers and their clients (40). It appears that coordinating 80 individuals becomes difficult at times especially when many of the home mangers do not have proper communication facility like phones. At present mobile ownership amongst the home managers is still limited. One innovation could be to provide mobile phones to the Home Managers so that the communication is facilitated. The mobile could be packaged into the training program and after the placement; the trainees can pay in easy monthly installments. As the demand for Home Managers increases, the coordination problem will increase unless addressed immediately.

Stakeholder Satisfaction

Discussion with community and various SAATH partner's indicate they hold the organization in high esteem. The community certainly appreciates the work of the organization because their interventions have increased access to basic facilities like water, sanitation, health and education. The income earning opportunities have substantially improved. The community feels empowered that they are heard by the AMC officials and can bargain and negotiate for services due to them.

The Government of Gujarat and its agency like Gujarat Urban Development Company and Gujarat Urban Development Mission and other institutions like the Ahmedabad Municipal Corporation have included SAATH most of their urban policy and project initiatives including the formulation of Gujarat Slum Policy and the Gujarat Urban Poverty Program. SAATH in considered as one of the most important organization having strong grass roots experiences which is flexible in its approach and can collaborate the government initiative. For the local authority, SAATH has provided new innovations and bridges to communicate with the community enhancing their image and legitimacy amongst the urban poor. The fact that the Government of Gujarat is expanding SAATH's initiative in other cities and regions, including tribal areas of the state is a significant achievement.

SAATH's initiative is also seen positively by the funding organization. Their involvement in the SNP has received World Bank appreciation and CARE is interested to support future SNP projects with SAATH. The URC and Umeed have also received positive response. The only weakness some funding organization highlight is that SAATH is weak in terms of research, documentation and reporting. The urban capability of the staff have also been identified as one of their weak areas the capacity for which needs to be strengthened.

Smaller NGOs like Parivartan appreciate the role of SAATH in providing them technical support for capacity building. The flexible and open approach has facilitated partnership building. However, to say this is true with all is not right. SAATH did have difference with Sharda Trust in Sanjayanagar while implementing the SNP. SAATH has not worked together with other prominent NGOs like MHT-a sister organization of SEWA, except may be for advocacy work. This however does not imply that they have to work together. Given the nature of development, especially service delivery, work the efficiency of respective organization lies rather in maintaining a balanced and strategic distance. Involvement of too many organizations may lead to conflicts that undermine the impacts.

Conclusions

As a part of INTRAC research the present author appraised SAATH and its activities in 1996. At that time, SAATH was in its formative stage and was struggling with the first few SNP projects in Ahmedabad. Most of its work was related to community mobilization and ISDP. Since then the organization appears to have grown substantially in terms of its reach, activities and innovations. Today it is recognized as one of the few important urban NGOs in the city of Ahmedabad and the State of Gujarat. Its legitimacy is high with the people with whom it works and also with the government, private sector, academics and other civil society institutions. It has grown in stature through its involvement in innovative social development concepts that includes the SNP, Umeed, the URC and RCH. It has scaled up its activities but has not compromised on building and empowering the poor and the marginalized. Many of the CBOs that were being formed in 1996 have matured and are taking up many responsibilities with minimum support from SAATH. Their activities have diversified beyond health and education to micro finance, livelihood and advocacy. New institutions like the URC is being managed and run by the CBOs. Many of these CBOs are fighting communalism and are proactively engaged in forming new CBOs. All these activities have made a difference not only in the lives of the community but also in terms of initiating new governance initiatives from the grass roots. In this reform process, SAATH has been able to leverage and build relationship with various stakeholders. It has fostered partnership models that are relevant and important in the liberalized and globalized city economy. Such partnership initiatives have brought institutions and people together. The difference and distance that used to exist between AMC and the slum dwellers have reduced as the scope for interaction has increased. Partnership and collaborative initiatives are also important in leveraging benefits of the market economy. Unlike many other organization SAATH is not shy of market forces but has succeeded and demonstrated, through its innovations, that market can benefit the people. In doing so SAATH has helped people to believe in their potential and creativity. The will and ambition to work hard and succeeded is amply reflected amongst the home mangers and Umeed trainees. All in all the initiative of SAATH since 1996 has consolidated and is certainly making a difference in the lives of the poor people. There is no mistake in making this assessment. The positive assessment however does not imply that there are no problems. SAATH is encountering all the problems that are true for any development organization. What is however important is that despite these problems SAATH is pro active and positive in search of new innovations and ideas to overcome these obstacles. The persistence to look for new ideas has made SAATH different from other organizations and this should also be the basis for its future growth.

Suggestions

- Long term financial planning and security is urgent requirement. SAATH should think of graduating from short-term project funding to long term program funding and a corpus needs to be created so that projects are not affected by financial constraints.

- Strengthen manpower base, if required professionalize. Second rung leadership is weak and needs to be developed.
- Synergize long-term financial, project and human resources planning. In absence of such a synergy, initiatives become reactive and ad hoc.
- Urban livelihood should be the core competency of SAATH around which other activities can be built. SAATH should not jump into many activities in the guise of innovation. It needs to be selective based on its organizational strength and competency.
- Participatory organizational assessment is required in the context of SAATH's emergence as an important urban NGO and also in the context of the rapidly changing nature of the city. Vision, mission and objectives need to be reviewed and if required reformulated. If necessary, a phased restructuring plan also should be worked out.
- SAATH's strength is in implementing and innovating projects at the grassroots. This should be continued and further strengthened. Research and policy activism should be supplementary not the core function otherwise it will dilute SAATH's grassroots competency.
- Evolve proper MIS system to facilitate decision-making process. Install internet connection at all field and sector office so that connectivity and decision making becomes easy and efficient. This is important in the context of rapid spatial and sectoral expansion of SAATH's work.

BIBLIOGRAPHY

1. Acharya, S. and L. Thomas, 1999, Finding a Pathway: Understanding the Work and Performance of NGOs in Ahmedabad, India, INTRAC, Oxford, U.K.
2. Casroline, S. 1995, Strengthening the Capacity of NGOs; Cases of Small Enterprise Development Agencies in Africa, INTRAC, Oxford, U.K.
3. SAATH Annual Reports, 2003–05, 2004–05, 2005–06 and Draft 2006–07, Ahmedabad.
4. SAATH occasional unpublished reports on URC, Savings and Credit, Home Managers (undated)

Appendix I

Diversity Chart

PARTICULARS	STAFF ENGAGED IN URBAN PROJECTS OF SAATH				
	HINDU DALIT	HINDU NON DALIT	MUSLIMS	TOTAL	%
MALE:					
Staff	10	28	7	45	
Volunteers	0	0	0	0	
SUB TOTAL	10	28	7	45	9
%	22	62	16	100	
FEMALE:					
Staff	22	25	16	63	
Volunteers	331	20	39	390	
SUB TOTAL	353	45	55	453	91
%	78	10	12	100	
GRAND TOTAL	363	73	62	498	
%	73	15	12	100	100

Appendix II

Saath Coverage in Ahmedabad City

Ward	Number of Slums	Activity	Beneficiaries/ Participants/ Members	House Holds
Amraiwadi	6	ICDS	411	
	1	SNP		210
Bhaipura Hatk	4	ICDS	234	
Bapunagar	20	RCH & JD		5,840
	1	ICDS	52	
Behrampura	1	Umeed	1,190	
	37	ICDS	2,077	
	6	MFI	3,849	
	4	CBO		3,631
	3	URC		1,332
	6	CVU		3,200
Bhaipura Hatkeshwer	2	ICDS	121	
	1	SNP		85
Danilimda	2	ICDS	105	
Dudheshwer	3	URC	500	
Gandhigram	3	ICDS	182	
Girdharnagar	1	ICDS	55	
Isanpur	1	ICDS	45	
Jamalpur	3	ICDS	170	
Juhapura	8	TB	40	
	3	Balghar	100	
	1	CBO		3,000
	3	URC		720
	5	CVU		1,010
	5	MFI	766	

Contd...

Ward	Number of Slums	Activity	Beneficiaries/ Participants/ Members	House Holds
Juna Vadaj	1	Umeed	167	
	10	ICDS	599	
Kankaria	28	ICDS	1,669	
	29	RCH & JD		4,533
	1	MFI	4	
Khadia	2	ICDS	113	
Khokhra	2	ICDS	115	
Madhupura	1	ICDS	75	
Maninagar	18	RCH & JD		5,118
	10	ICDS	661	
Naroda	4	ICDS	253	
Naroda muthia	2	ICDS	117	
Nava Vadaj	1	Umeed	179	
Nikol	1	ICDS	35	
Odhav	8	ICDS	532	
	1	SNP		109
Paldi	7	RCH		3,567
Potalia	3	ICDS	220	
Raikhad	3	ICDS	161	
Rajpur	16	ICDS	926	
Rakhial	2	ICDS	146	
Sabarmati	4	ICDS	197	
Sardar Patel Stadium	1	SNP		170
	10	ICDS	582	
Saraspur	34	RCH & JD		5,432
Sardarnagr	6	ICDS	439	
	6	MFI	698	
	1	CBO		1,250
	6	CVU		1,500
	2	ICDS	105	

Ward	Number of Slums	Activity	Beneficiaries/ Participants/ Members	House Holds
Sarkhej	1	Home Managar	41	
Shahpur	1	ICDS	63	
Vadaj	1	URC		250
	6	CVU		1,150
	1	ICDS	48	
Vasna	44	RCH		10,024
Vatva	63	RCH & JD		7,737
	1	SNP		545
	1	CBO		1,350
	2	Home Manager	69	
	1	URC		8,906
	5	CVU		1,250
	3	House Keeping	14	
	5	MFI	3,194	
	14	ICDS	926	
Vejalpur	1	Home Manager	22	
Totals	**500**		**22,267**	**71,919**

www.ingramcontent.com/pod-product-compliance
Lightning Source LLC
Chambersburg PA
CBHW020852180526
45163CB00007B/2484